EXPLORING ELECTRICITY & ELECTRONICS
WITH PROJECTS

BY JOHN EDWARDS

TAB TAB BOOKS Inc.
BLUE RIDGE SUMMIT, PA. 17214

FIRST EDITION

FIRST PRINTING

Copyright © 1983 by TAB BOOKS Inc.

Printed in the United States of America

Reproduction or publication of the content in any manner, without express permission of the publisher, is prohibited. No liability is assumed with respect to the use of the information herein.

Library of Congress Cataloging in Publication Data

Edwards, John, 1954-
    Exploring electricity and electronics with
projects.

    Includes index.
    1. Electronics—Amateurs' manuals. I. Title.
TK9965.E3   1983   621.3   82-19442
ISBN 0-8306-0497-9
ISBN 0-8306-1497-4 (pbk.)
Cover illustration by Al Cozzi.

Contents

To Mom, for enduring

Acknowledgments

I wish to extend my deep appreciation and thanks to the following organizations for their generous help in the preparation of this book:

Duracell International, Inc.
Electronic Specialists, Inc.
Hewlett-Packard
National Aeronautics and Space Administration
Panasonic
Radio Shack
Sprague Electric Company
Texas Instruments, Inc.
William M. Nye Company, Inc.
Yaesu Electronics Corporation

The most beautiful thing we can experience is the mysterious. It is the source of all true art and science.

Albert Einstein, *What I Believe*

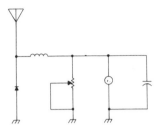

Introduction

Why are you reading this book? Let's be honest, there are more exciting things to do with your time—watching TV, for instance. Of course, television viewing and many other late twentieth century pastimes depend on the subject of this book for their existence. So while the solitary act of reading may not send chills up and down your spine, reading this book can help lead you to a better understanding of those gadgets that make life today so interesting.

The dictionary defines electronics as the branch of physics concerned with the study and control of electrons, but there's more to it than that. Electronics has taken humanity to the moon and beyond; it makes possible television, radio, pocket calculators, and the thousands of other astounding devices we all take for granted.

Increasingly, electronics has become a tool—the space age hammer. With a knowledge of electronics, one can modify his environment to make life more comfortable and productive. Think of this work as an instruction book for this tool. Use it well.

During the course of this book, you may want to try your hand at building one or more of the included projects. Here's a few hints for the beginning electronics hobbyist. The parts listed are those most commonly available, but remember that many parts numbers are interchangeable. Various companies have different numbers for the same part. Therefore, a good substitution handbook is a must. If you don't want to buy one, your local library or TV service shop may have one you can borrow. Also, for many less critical parts (resistors, capacitors, etc.), you can "cheat" on values by as much as 20 percent without adversely affecting the circuit's performance. You're an electronics experimenter, so be sure to experiment.

Within recent years, finding parts has become an increasingly difficult task. While designing projects for this book, great care has been taken in selecting parts that are readily available from retail electronics stores. Nevertheless, even if you live in a major metropolitan area, you will probably have to resort to mail-order suppliers to find at least some parts. A good way to select mail-order parts suppliers is through ads in electronics magazines. Don't bother writing to manufacturers for parts. Virtually all manufacturers sell only in bulk volume to electronics jobbers. All in all, just consider this another facet of the challenge of electronics.

Another consideration in designing projects for this book has been the actual physical layout of each project board. While printed circuit boards are now standard for most solid-state projects, I've decided to go the breadboard route. This is because making printed-circuit boards can be a harrowing experience for the beginner and this book concentrates on teaching electronics fundamentals, not the basics of etching baths and the like.

One final note on the art of soldering. Ninety-nine times out of a hundred, when a beginner's project fails to work, the problem can be traced to faulty soldering. If you're not confident of your soldering skills, try to find someone willing to teach you. There's a knack to this, and it's next to impossible to learn to solder correctly by just reading a book. In general, remember to keep your iron's tip clean, preheat the component leads *before* applying the solder, and use only enough solder to assure a good electrical connection.

Now, let's get down to business.

Chapter 1

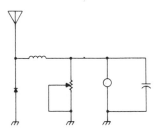

Electricity and Its Sources

When you take everything in electronics into account—all of the radios, guidance control systems, test equipment, radar scopes, and other hardware that affects our daily lives—it all boils down to one simple little particle of our universe: the electron.

The electron is everything and nothing. In scientific terms, it has a mass of 9.107×10^{-28} gram—so small, that even with today's fantastic technology it has only barely been glimpsed with an electron microscope. Yet, this magnificent nothing controls everything in electronics from the tiniest calculator to the mightiest computer system.

How can such an insignificant bit of matter exert such a force? The answer lies in the electron's ability to hold a negative electric charge. While orbiting around a positively charged group of protons, the electron remains in a nominally neutral state. But disturb this relationship, set the electrons in motion by moving them from one place to another, and you create a flow of electrons—*electricity*. Figure 1-1 shows a simplified diagram of the atomic structure.

VOLTAGE

Since all matter is composed of atoms (containing both protons and electrons), all matter must be in one of three states: positively, negatively, or neutrally charged. When an object is positively charged, it has more protons than electrons. An object that is negatively charged has more electrons than protons. A neutrally charged piece of matter contains equal numbers of protons and electrons.

Nature, on the whole, dislikes imbalances. Therefore, positive and negative charges tend to cancel each other. If you bring a negatively

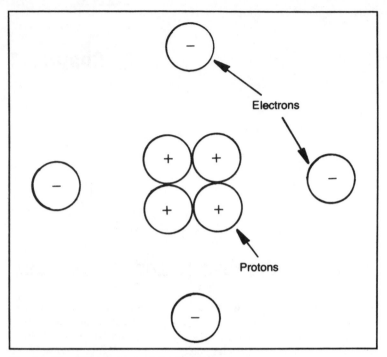

Fig. 1-1. Simplified diagram of atomic structure.

charged bit of matter near something that is positively charged, the charges will try to switch between the atoms. But, as seen in Fig. 1-2, only electrons have the ability to make the switch. Except in rare instances (such as in atomic explosions), protons remain within the atom's nucleus—that's why this isn't a book on "protonics."

The force with which electrons travel is called the *difference of potential* or, in its more common term, *voltage*. In electronics, voltage is the stuff that makes all our gadgets run. This "push" or, as it's sometimes also called, *electromotive force,* can be produced by a number of different methods. Chemical action, electrical generation, heat, electrostatic action, magneto-hydrodynamic action, and photo and piezoelectric effects are all sources. Don't let the strange words overwhelm you, however, we'll look into the most common sources of voltage in a little while.

CURRENT

When one hears the phrase "a stream of electricity," the imagery really isn't far from the truth. In many respects, a flow of electrons does behave very much like water in a stream. If one can imagine an electron being like a molecule of water, it may make understanding the basics of electronics easier.

2

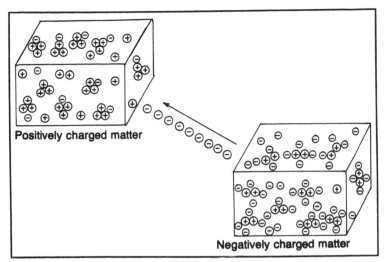

Fig. 1-2. Electrons flow from negatively to positively charged matter.

Like a stream, electricity has a force to its flow. As we've just learned, this force is called *voltage* and is measured in *volts*. And, like water, electricity also flows at a set rate; this we call *current*, which is measured in *amperes*. Sometimes, the difference between voltage and current is a bit hard for the newcomer to grasp, so let's take a closer look at the two.

Figure 1-3 shows two meters—an "electrical pressure meter (A)," and a "flow meter (B)." The pressure meter measures the force behind the electron flow: the voltage. The flow meter's task is to see how many

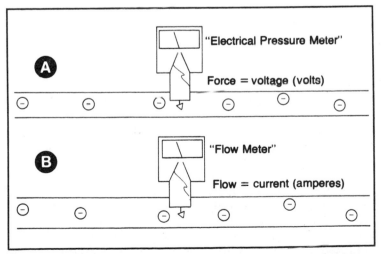

Fig. 1-3. Measuring force and flow: two very different aspects of electricity.

electrons are passing a point in a given amount of time: the current. Actually, there are two meters that do these very jobs. We'll learn more about them in the chapter on test equipment. The point to remember for now is that voltage and current are two different, (and very important) aspects of electricity.

BATTERIES

As we've pointed out, electricity and flowing water bear many similar characteristics. One other way these two natural forces resemble each other is in the methods used to generate their flows. To make water move, one might use a water pump. Likewise, in electronics, we use our own version of the pump—the chemical action, electrical generation and other methods used to produce voltage we mentioned earlier. Out of all of these generation techniques, one of the most common in day-to-day electronics use is chemical action. And when we talk about chemical electrical production, we're generally speaking about the *battery*.

The theory behind battery operation is quite simple. The battery's single task is to provide a flow of electrons. This is accomplished by a chemical reaction within the battery. Each battery has two *electrodes* (metal rods); one positive, the other negative. Each electrode is made of a metal dissimilar to the opposing electrode. Surrounding the electrodes is a substance called an *electrolyte*. The electrolyte acts as a medium, carrying electrons from the negative electrode (which has an abundance of electrons) to the positive electrode (which is hungry for electrons). This movement of electrons then gives us our flow (current) and force (voltage): our electricity stream.

Batteries come in many shapes and sizes, depending on the job that's needed to be done. Batteries themselves are made up of individual chemical units called *cells*. Each of these cells contain the electrolyte and two electrodes described above. The most popular cell for ordinary use, because of its relatively compact size and low cost, is the *carbon cell*. Figure 1-4 shows a conventional carbon cell.

Carbon Cells

Carbon cells get their name from the carbon rod they use as a positive terminal. Were you to slice open a carbon cell (not recommended for safety reasons), you would see that this rod is surrounded by an electrolyte made from a powdery mixture of ammonium chloride, zinc chloride and other materials. The case, which acts as a negative terminal, is made out of zinc. Figure 1-5 shows a cross-section of carbon cell.

The cell shown in Fig. 1-4 is a common D-type cell. As you are probably aware, carbon cells are available in a variety of sizes. Small AA penlights, medium-sized C cells and larger D units are only a few of the many popular sizes. Strangely enough, however, as you increase the size of a carbon cell, its voltage remains the same. Therefore, AA and D cells each

Fig. 1-4. The carbon cell.

produce the same voltage—1.5 volts—regardless of their respective sizes. But what if a device requires more than 1.5 volts for operation? Sure, you could always use more than one cell. But sometimes, especially outdoors, using multiple cells can be, at best, inconvenient. What's needed is a series of carbon cells in one compact unit.

Only when a series of cells are connected together to produce a device capable of generating higher voltage, do we have a true battery. Unfortunately, the terms "battery" and "cell" have become almost synonymous, leading to much unnecessary confusion. Just as many people might incorrectly call a D cell a battery, most people would probably refer to the item in Fig. 1-6 as a "dry cell." While the lantern battery in this photograph contains a group of dry cells harnessed together (in this case, four carbon cells to

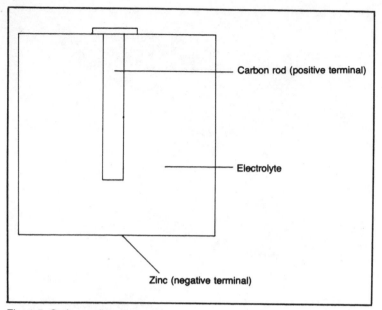

Fig. 1-5. Carbon cell cross-section.

produce six volts), collectively they are called a battery, not a cell. Understand? We hope so. Even some electrical engineers have been guilty of confusing these terms.

Other Dry Cells

Although carbon cells are quite popular, they do have a number of drawbacks. Compared to other members of the dry cell family (called "dry," because of their non-liquid electrolytes), carbon cells generally store less energy and have a shorter life. Of course, the carbon cell's low price works in its favor, but cost can be a relatively minor consideration when reliable service is important. In response to these problems, technology has come up with a whole series of cells designed for different applications. While some of these units may resemble carbon cells in appearance, the chemistry inside these devices functions in an entirely different manner.

Alkaline Cells

Perhaps the most common alternative to the carbon cell is the *alkaline cell* (Fig. 1-7). Manufactured in cases that look similar to those of carbons, you're probably most familiar with these cells from their advertiser's claims of longer life. For once, the advertisers are correct. Alkalines, under the right conditions, last considerably longer than an equivalent size carbon cell. Of course, alkalines can cost up to twice as much as carbons, so their main effectiveness lies in situations where reliability is important.

Internally, alkalines have a much more sophisticated innerstructure than carbons. But unless you're thinking of making a career in the battery field, the alkaline cell's actual construction isn't really very important. It is important to remember, again, that alkalines possess a longer working and shelf life than carbon cells. For a closer look at the alkaline cell see Fig. 1-8.

Nickel-Cadmium Cells

So far, the cells we've been discussing are categorized as *primary cells*. That is, once their original charge has been exhausted (after all the electrons have been transferred to the negative electrode), they cannot be returned to their original state. But what if we have a battery-operated appliance that we use daily over a long period of time? Do we buy new cells

Fig. 1-6. This lantern battery consists of four individual carbon cells.

7

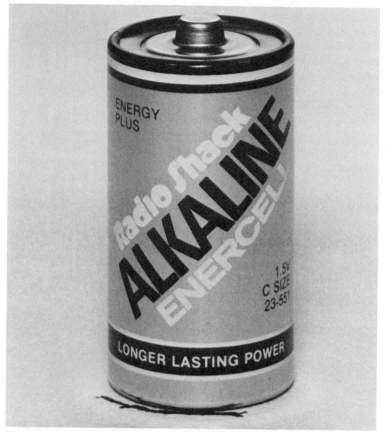

Fig. 1-7. The alkaline cell.

each time the old ones wear out? That would get awfully expensive—and wasteful. To fill the need for a cell that can be "recycled," Science has devised the *secondary cell*.

Secondary cells, as their name implies, can be charged a second time. As a matter of fact, they can be charged a third time, fourth time—up to a thousand charges, in many cases. While a number of secondary cells exist—some very experimental—the most popular type is the *nickel-cadmium cell* (Fig. 1-9).

Nickel-cadmium cells (nicads, as they are more commonly called) are used in all sorts of applications where repeated charges are required. Hand-held police radios, hearing aids, and electronic wristwatches are only a few of the places you'll find nicads. Unlike most other dry cells, nicads can be recharged to an almost-new condition within a matter of hours with no damage to the cell (Fig. 1-10). Carbon and alkaline cells, on the other hand, have the nasty tendency to leak or explode during recharging.

But like all good things, nicads have their bad side, too. Their cost is almost outrageous (as much as four times that of an alkaline cell), they have a very limited shelf life, and even at full charge they hold only about 25% as much energy as an alkaline. All of these factors have limited the nickel-cadmium cell to a very small share of the consumer market.

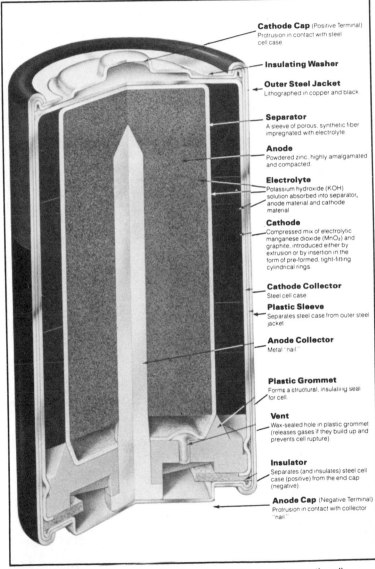

Cathode Cap (Positive Terminal)
Protrusion in contact with steel cell case

Insulating Washer

Outer Steel Jacket
Lithographed in copper and black

Separator
A sleeve of porous, synthetic fiber impregnated with electrolyte

Anode
Powdered zinc, highly amalgamated and compacted

Electrolyte
Potassium hydroxide (KOH) solution absorbed into separator, anode material and cathode material

Cathode
Compressed mix of electrolytic manganese dioxide (MnO_2) and graphite, introduced either by extrusion or by insertion in the form of pre-formed, tight-fitting cylindrical rings

Cathode Collector
Steel cell case

Plastic Sleeve
Separates steel case from outer steel jacket

Anode Collector
Metal "nail"

Plastic Grommet
Forms a structural, insulating seal for cell.

Vent
Wax-sealed hole in plastic grommet (releases gases if they build up and prevents cell rupture).

Insulator
Separates (and insulates) steel cell case (positive) from the end cap (negative).

Anode Cap (Negative Terminal)
Protrusion in contact with collector "nail"

Fig. 1-8. Alkaline cell cross-section (courtesy of Duracell International).

9

Fig. 1-9. Some different nicad types.

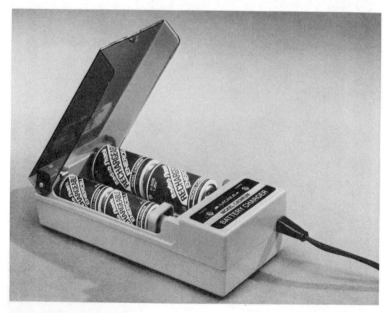

Fig. 1-10. The nicad's main advantage over carbons and alkalines, is that it can be recharged.

Wet Cells

There's an entire group of batteries we've managed to ignore so far—the wet cell family. While their use is very limited in hobby electronics, it still pays to know a little bit about them. Although you probably don't have a wet cell in your home, there's a good chance you have one either in your garage or parked outside your front door, since all automobiles use a wet cell battery to power their electrical systems. The wet cell battery found in cars, and the type in most common use today, is the *lead-acid cell* (Fig. 1-11).

Lead-acid cells have an electrolyte formed from a mixture of sulphuric acid and water. Rods of lead peroxide and sponge lead form the positive and negative electrodes, respectively. With lead electrodes and a liquid electrolyte, lead-acid batteries, as you might imagine, weigh a proverbial ton. Needless to say, this doesn't make them very practical for pocket radios or calculators.

It's really quite unfortunate that lead-acid batteries are so bulky. Being secondary cells, they can be recharged many times and are capable of producing a great deal of power. But even though they do have some mighty drawbacks, many uses are found for wet cells in the transportation field, in places where commercial power is limited, and as back-up power supplies when commercial power fails.

ELECTRICITY FROM MAGNETISM

The electricity produced by batteries works very well, but battery power itself has many disadvantages. As we've noted, all batteries, no matter how well designed, eventually wear out. Also, the amount of power a battery can deliver is severely limited. Obviously, if batteries could supply all of our energy requirements, there wouldn't be any need for electric companies to run power lines into our homes—we would just operate all of

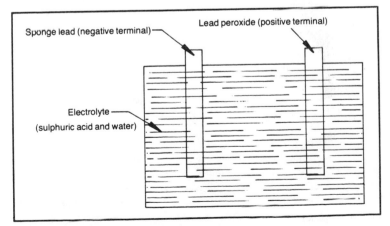

Fig. 1-11. Lead-acid cell cross-section.

our appliances from a series of cells. But, sadly, this is not the case in the real world. Batteries just aren't adequate to power large electrical devices at a reasonable cost.

To circumvent the battery's shortcomings, devices known as *generators* have been developed to supply our electric needs. While few of us own generators in the home (except perhaps for small ones used on camping trips or in case of "blackouts"), electric companies make extensive use of these devices to light our homes, streets, and businesses.

As seen in Fig. 1-12, generators function through the use of magnetism. While, to most of us, magnets are nothing but a child's toy or something we use to attach messages to a metallic bulletin board, magnets and magnetism play a crucial role in electronics. Throughout the course of this book, magnetism will keep popping up in different forms and guises, but for now let's see how magnetism helps us generate electric power.

Generator Mechanics

Referring again to Fig. 1-12, you'll see that a generator basically consists of a *wire loop* suspended between two oppositely polarized magnets. This general structure applies to any generator, whether it's of the backyard-camping variety, or if it powers half a city. The wire loop is mounted on a revolving core called an *armature*. The armature is powered from an external source such as coal, oil, geothermal, or hydroelectric power. It's this need to turn the armature that makes generated electric power so dependent on outside energy sources.

As the armature spins around, its metal loop is cutting through the field created by the two magnets. Current is transferred out of the generator via a pair of *slip rings* mounted around the armature. The slip rings are in constant contact with a stationary set of *carbon brushes* that attract the current from the rings and move it into the outside circuit.

In practice, a generator will actually have a great many wire loops (wound in the form of coils) and an entire series of magnetic poles for the loops to play off of. But, for illustrative purposes, we've tried to keep things at the simplest level. The theory, nevertheless, is accurate.

Ac and Dc

Now, all of this seems pretty magical. After all, how can a loop of wire merely spinning through a magnetic field produce power? Well, in many respects it is magical—as any natural phenomenon is—but there's also a way of explaining the process.

Whenever a wire loop is rotated between the poles of a magnet, voltage will be induced within that wire. This occurs when the wire cuts through the edge of a magnetic field at a high speed (sixty times per second in conventional home power). Basically, all a generator does is to convert one type of energy into another. The coal, oil, or water power that is used to spin the armature is converted, through the help of the magnets, into electric power.

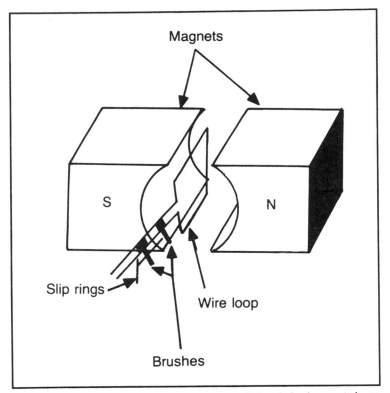

Fig. 1-12. A wire loop spinning between two oppositely polarized magnets forms the heart of an electrical generator.

The power created by a generator is different from that of a battery. Batteries produce what is called *direct current,* or *dc.* Generators, on the other hand, create *alternating current,* or *ac.* If you'll remember back to the beginning of this chapter, you'll recall that we said electricity behaves like a stream. But, unlike earthly rivers, electrical streams can sometimes reverse themselves and flow in the opposite direction. Current that flows in a single direction is dc; current that alternates its flow forwards and backwards is ac.

Each time the wire loop in a generator makes one complete turn (as shown in Fig. 1-13), its output moves from zero, to a high level in one direction, falls to zero again, back to high in an opposite direction, and then down to zero again. If you were to plot this output graphically, calling one direction positive and the other negative, it would appear as in Fig. 1-14. We call this representation a *sine wave.* As you can see, during the first half of the cycle, the current flows in the positive direction; during the final cycle half, the output falls into the negative zone. Ordinary household ac, which completes sixty such cycles during the course of a second, is said to operate

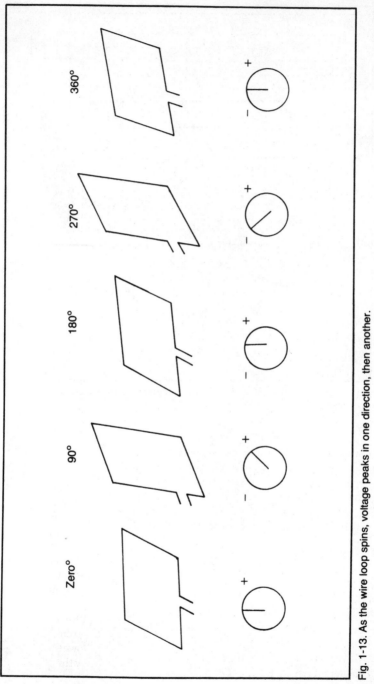

Fig. 1-13. As the wire loop spins, voltage peaks in one direction, then another.

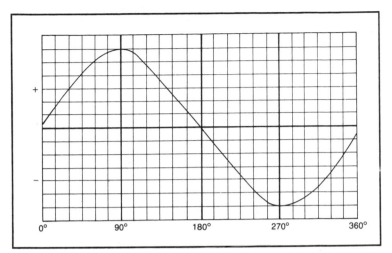

Fig. 1-14. A graphic representation of ac: the sine wave.

at a *frequency* of 60 *hertz* (hertz, being the official designation for cycles-per-second).

OTHER SOURCES OF ELECTRICITY

Battery and generator power are the two most common ways to produce electricity, but other sources are available. The problem is that, for the most part, the electricity generated by these other sources is either very weak, difficult to harness, very expensive, or a combination of all of the above. As long as you realize that batteries and generation are the two most common methods of electrical production, however, you should have nothing to worry about.

PROJECT 1: ELECTRICAL CHARGE

An electron flow is, of course, invisible. But there are definite ways of seeing this flow at work. One way, is through this little project.

Figure 1-15 shows the simple household items we can use for a visual demonstration of electrical flow. Merely take a piece of tissue paper and tear from it three or four small pieces (about ½" by ½"). Next, take an ordinary comb and rub it with a flannel cloth (those cloths used to clean phonograph records seem to work especially well). Carefully move the comb close to the pieces of paper, and they should suddenly jump onto the comb.

What happened? Well, when you rubbed the comb with the cloth, you gave the comb a negative electrical charge—a load of electrons. Since the paper is neutrally charged, the electrons within the comb flowed toward the paper (nature dislikes imbalances, remember?), which drew the paper onto the comb.

15

Now, carefully remove the paper from the comb with a pencil. With the paper back on the table, bring the comb once again close to the tissue fragments. This time, instead of leaping onto the comb, they should remain on the table.

While they were on the comb, the paper absorbed the comb's negative charge. Since the paper now has an equal electrical charge, the tissue is no longer attracted.

A note: this experiment is not always completely reliable. Since we are dealing with extremely weak charges, many environmental factors may interfere with this demonstration. On the whole, things work best on a dry, cool day. You may also wish to try a variety of different combs, since some seem to work better than others.

PROJECT 2: POWER FROM THE SUN

In this chapter. we've concentrated on batteries and generators as the two primary means of creating electricity. Yet, as we've mentioned, other methods do exist. And perhaps no other mode offers as great a promise as solar power.

As this book is being written, solar energy is still in its infancy, its applications limited to a relative handful of uses. However, the potential of harnessing the power of the sun gives one hope that someday we can be freed from the limits of such present-day energy sources as coal and oil.

As detailed in Fig. 1-16, the aim of this project is to convert raw solar energy into electricity through a solar panel, using it to drive a simple buzzer. Since solar energy is variable (clouds, buildings, etc., may block sunlight), the amount of voltage and current sent to the buzzer will also vary. This will cause the buzzer to change from a low to high pitch as you move from shadowed areas into direct sunlight.

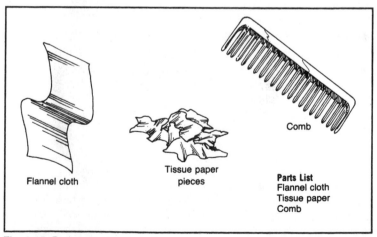

Flannel cloth

Tissue paper
pieces

Comb

Parts List
Flannel cloth
Tissue paper
Comb

Fig. 1-15. Project 1: electrical charge.

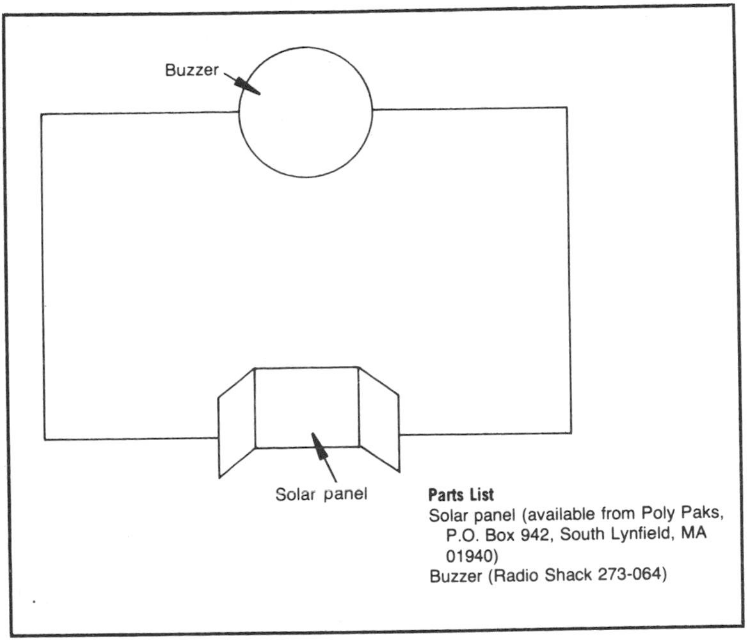
Buzzer
Solar panel
Parts List
Solar panel (available from Poly Paks, P.O. Box 942, South Lynfield, MA 01940)
Buzzer (Radio Shack 273-064)

Fig. 1-16. Project 2: solar-powered buzzer.

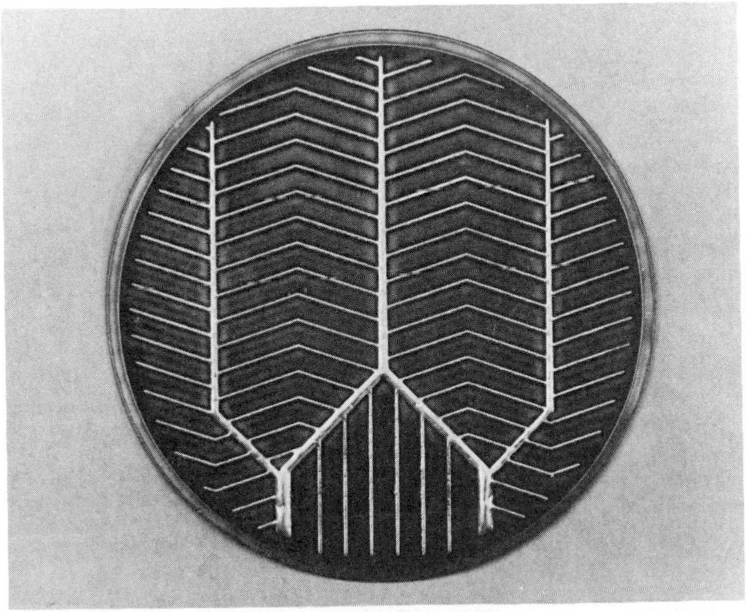

Fig. 1-17. A solar cell: connect a few in a row and you have a solar panel.

Actually, you may think that converting solar energy into power to run a buzzer isn't such a big deal—and it really isn't. Yet, as the ability of solar cells (Fig. 1-17) to convert sunshine into higher amounts of voltage and current increases, perhaps we may someday derive most of our household power from this source. And what would one do at night? Well, we could always store solar energy in batteries during daytime to use when it's dark.

Give this project a try. It's very simple; just connect it together as shown in the diagram. You may be experimenting with mankind's future.

Chapter 2

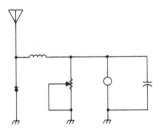

Direct-Current Circuits

In the previous chapter, we learned what electricity is and how it can be produced. But electricity all by itself isn't very useful—a flow of electrons leading nowhere can't do very much. What's needed is a way to harness this power and put it to work. That's the task of the direct-current (dc) circuit.

So far, we've discussed two different types of electrical energy: direct current (dc) and alternating current (ac). As you may suspect, one can only use direct current within a dc circuit. This isn't to say that ac doesn't have its uses—it does—but we'll look into them in later chapters. For now, we're going to concentrate on the important world of dc circuitry.

THE CIRCUIT

In order for a circuit to be complete, it must consist of at least three different elements: a power source, a load (a device to be operated from the source), and pathways to connect these two components together. Figure 2-1 shows an elementary circuit using a dry cell power source, a lamp for a load, and copper wire interconnecting the two. When the circuit is assembled, current flows from the negative battery terminal to the lamp and back to the battery's positive terminal. We have put our electrons to work by illuminating a lamp.

SCHEMATIC DIAGRAMS

Of course, this is a very simple circuit. Other circuits have been devised that include thousands of different components connected in very intricate patterns. Nevertheless, no matter how complicated or simple a circuit may be, people need a way to follow and plot their construction. Otherwise, electronics would be a science without structure.

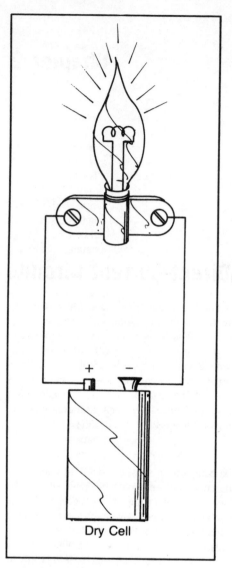

Fig. 2-1. Elementary dc circuit.

Dry Cell

In many respects, a circuit may be looked upon as a sort of electrical highway system. Roads lead in all directions, electrons traveling along them may pause or terminate at various spots, others continue on their way. And like a highway system, engineers have developed a way to map the maze of electronic circuitry so that interested parties may trace their way along.

On this specialized road map—called a *schematic diagram*—all electronic components are given their own symbols so they can be identified at a moment's glance. Figure 2-2 shows some of the common symbols we'll be using during the course of this book, and Fig. 2-3 demonstrates how schematics can be applied, by illustrating our battery-lamp circuit in schematic form. Study both illustrations and you'll see that schematics are really quite easy to understand.

RESISTANCE

Returning to our analogy between electricity and flowing water for a moment, we've neglected to mention yet still another parallel between the two: obstructions that might hamper their flow. Many things can work to slow the course of a river: boulders in the stream, fallen branches, a rough river bottom, etc. Similarly, a stream of electricity can also be interfered with. We call this action *resistance*, and measure it in *ohms* (Ω).

Just as numerous things can slow a river's flow, electrical flow (current can be reduced in many ways, too. If the current is flowing through a wire, increasing the wire's length would slow it down. So would a wire's thickness and its material (a copper or silver wire has less resistance than, say, one made from iron or lead).

Resistance isn't always a nuisance, however. Imagine, for instance, a rampaging river without any dams to control its flow. Such a river, if left alone, would threaten many lives and property. In the same way, electrical resistance can be a force for good or bad. When resistance is unwanted (like a big rock in the middle of a busy waterway), it's a headache; when it functions as a means of controlling electrical flow (like a dam), it's very useful.

RESISTORS

The components we use to control electrical current are called *resistors*, and the shapes and forms these devices come in are almost as varied as their uses. Each resistor is specifically manufactured so that it adds an exact amount of resistance into a circuit—just enough for the control desired.

Most ordinary low power circuits use resistors made from a combination of carbon and clay. These are called *carbon resistors* (Figs. 2-4, 2-5), and are by far the most common type used. For high voltage applications, *wire wound resistors* are employed (Figs. 2-6, 2-7). These resistors contain lengths of highly resistant wire to introduce a specified resistance into a circuit. Yet another resistor type, used when circuit resistance must be altered to varying levels, is the *variable resistor* (Fig. 2-8). You're probably most familiar with this resistor type from its use as the volume control on your radio or TV.

RESISTOR COLOR CODE

For the most part, carbon resistors are pretty much indistinguishable from one another, regardless of their value. To help electronics designers, builders, and repairmen tell resistors apart, a color code has been developed to indicate their value and accuracy.

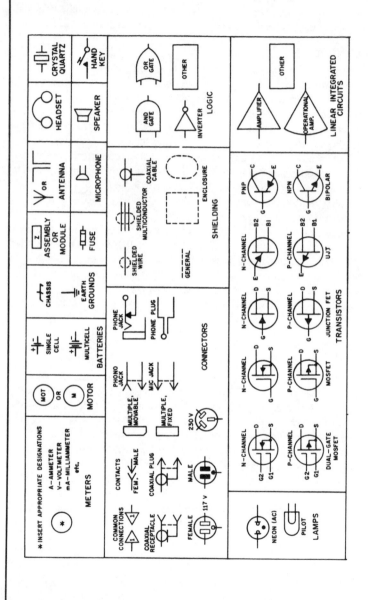

* INSERT APPROPRIATE DESIGNATIONS
A—AMMETER
V—VOLTMETER
mA—MILLIAMMETER
etc.

METERS

MOTOR

BATTERIES

SINGLE CELL

MULTICELL

CHASSIS

EARTH GROUNDS

ASSEMBLY OR MODULE

FUSE

ANTENNA

MICROPHONE

HEADSET

SPEAKER

CRYSTAL QUARTZ

HAND KEY

LOGIC

AND GATE

OR GATE

INVERTER

OTHER

LINEAR INTEGRATED CIRCUITS

AMPLIFIER

OPERATIONAL AMP.

OTHER

COMMON CONNECTIONS

CONTACTS

FEM.   MALE

COAXIAL RECEPTACLE

COAXIAL PLUG

FEMALE

MALE

117 V

230 V

PHONO JACK

MIC JACK

MULTIPLE, MOVABLE

MULTIPLE, FIXED

PHONE JACK

PHONE PLUG

CONNECTORS

SHIELDED WIRE

SHIELDED MULTICONDUCTOR

COAXIAL CABLE

GENERAL

ENCLOSURE

SHIELDING

NEON (AC)

PILOT

LAMPS

N—CHANNEL

P—CHANNEL

DUAL—GATE MOSFET

N—CHANNEL

P—CHANNEL

MOSFET

N—CHANNEL

P—CHANNEL

JUNCTION FET TRANSISTORS

N—CHANNEL

P—CHANNEL

UJT

PNP

NPN

BIPOLAR

22

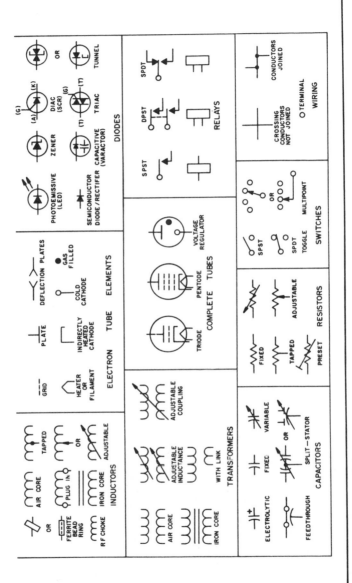

Fig. 2-2. Chart of schematic symbols.

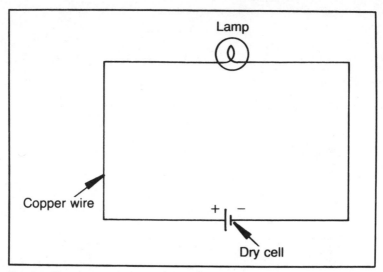

Fig. 2-3. Elementary dc circuit—in schematic form.

Looking at an ordinary carbon resistor, from left to right (Fig. 2-9), the first band indicates the first significant figure in the total resistance (in ohms) of the component. The second band denotes the second figure. The third band is a decimal multiplier, which tells the user how many zeros to add after the first two numbers. In other words, say a resistor has bands colored green, red, and red. According to the chart in Fig. 2-10, green

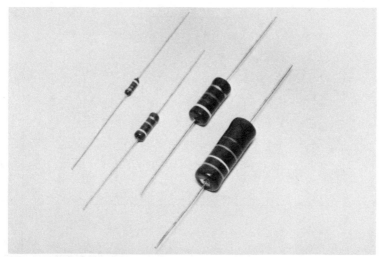

Fig. 2-4. Carbon resistors—the larger ones handle more power, they don't necessarily give greater resistance.

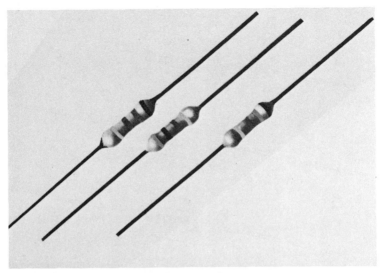

Fig. 2-5. Carbon resistors in another type of case.

stands for 5, red for 2, and the final red band a multiplier of 100. Therefore, the resistance offered is 5200 ohms.

Following the first three bands is a fourth gold or silver band. This is the resistor tolerance indicator: gold specifies a total resistance accuracy within 5%, silver 10%. Some cheap resistors have no fourth band. This indicates a tolerance factor of 20%. Non-carbon resistors, because of their larger size, usually have their values printed directly on their bodies.

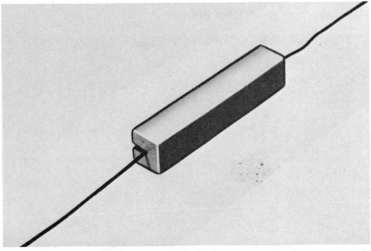

Fig. 2-6. Wire wound resistors—used in high voltage applications.

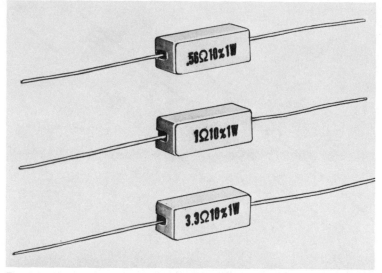

Fig. 2-7. Wire wound resistors have values printed directly on their bodies.

OHM'S LAW

Voltage, current, and resistance are the three basic foundations of electronics. Without these three factors, which must exist in every circuit, no electronic devices we make could ever possibly work.

As basic foundations, all three of these units are inexorably linked together. This is to say that if we are given a circuit wherein we know only the value of two out of these three factors, we can find the value of the third by following a simple mathematical formula. This principle is known as *Ohm's law*. Here are its three formulas:

Fig. 2-8. A variable resistor type — the potentiometer.

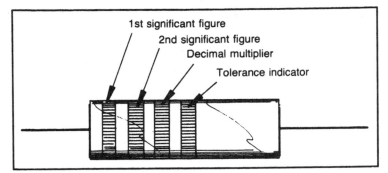

1st significant figure
2nd significant figure
Decimal multiplier
Tolerance indicator

Fig. 2-9. A carbon resistor's color code bands.

$$I = \frac{E}{R} \quad E = IR \quad R = \frac{E}{I}$$

In these formulas, "I" means current, "E" stands for voltage, and resistance is "R." Ohm's law is really quite an easy concept to follow once you get the hang of it, and its uses in electronics are legion.

USING OHM'S LAW

To help you leap over the Ohm's law hurdle, let's look at some ways we can use this principle. In Fig. 2-11, we see a simple circuit consisting of a battery delivering 1.5 volts at 3 amperes, connected to a resistor of unknown value. How can we determine its resistance? By using Ohm's law thusly:

$$R = \frac{E}{I} = \frac{1.5}{3} = .05$$

Color	Significant Figure	Decimal Multiplier	Tolerance
Black	0	1	-
Brown	1	10	-
Red	2	100	-
Orange	3	1,000	-
Yellow	4	10,000	-
Green	5	100,000	-
Blue	6	1,000,000	-
Violet	7	10,000,000	-
Gray	8	100,000,000	-
White	9	1,000,000,000	-
Gold	-	-	5%
Silver	-	-	10%

Fig. 2-10. Resistor color code chart.

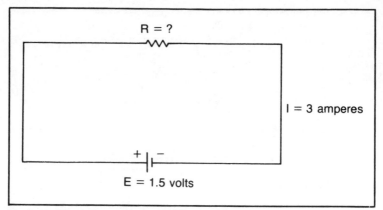

Fig. 2-11. What's the resistance in this circuit?

Since resistance must equal voltage divided by current, and in this circuit 1.5 volts divided by 3 amperes equals .5, the answer must be .5 or one-half ohm.

Likewise, in Fig. 2-12, we have another circuit using similar components; only this time we must deduce the circuit's current. Again, plugging the appropriate numbers into their correct slots, we come up with this:

$$I = \frac{E}{R} = \frac{1}{10} = .1$$

The answer, of course, is .1 ampere.

Finally, in Fig. 2-13, we must determine the circuit's voltage:

$$E = I \times R = 2 \times 10 = 20$$

Therefore, the voltage must be 20 volts.

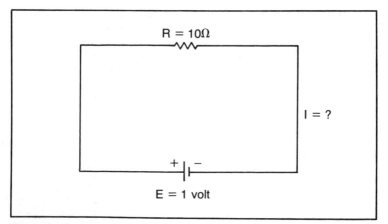

Fig. 2-12. What's the current?

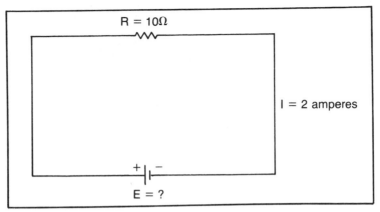

Fig. 2-13. What's the voltage?

Try making up some Ohm's law problems of your own. After a little practice, you'll become a real expert on the subject.

POWER

While we're on the subject of formulas, there's one more important principle that we should know: the factor of *power*. In electronics, power represents the rate at which energy is supplied to a circuit, or the rate at which it is consumed. Put another way, power is the unit of the rate of doing work, or producing energy. Power is measured in *watts* and has the symbol "P."

Like Ohm's law, the power formula works with voltage, current, and resistance. From a circuit's voltage and current, we can determine its power. Or, by knowing a circuit's power and voltage, we can figure out its current. Then, by using Ohm's law, we can find its resistance. It works like this:

$$P = EI, \quad I = \frac{P}{E}, \quad R = \frac{E}{I}$$

So, let's say we have a 7.5 watt lamp running at 110 volts. How much current is it drawing? Using our formula: 0.0682 amperes. How much resistance does the lamp and circuit offer? 1612 ohms. Easy as cake.

Uses of the power unit are all around us. As we've just seen, lamps are rated in watts, so are radio transmitter outputs, heaters, and many other appliances. Electric companies even bill us on the basis of how many watts we use in an hour. Power is indeed all around us, and knowing how to find and measure it can be very handy.

RESISTORS IN CIRCUITS

Now that we have a handle on resistors and what they can do, let's try putting them into a circuit. In our problems on Ohm's law, we saw how to place a single resistor into a circuit. But what if we have a circuit that needs more than a single resistor? How do we insert these components, and how much total resistance would they offer?

Series Circuit

When resistors are joined end to end, as in Fig. 2-14, we say they are connected in a *series circuit*. In such a circuit, the total resistor value is equal to the sum of the individual resistors. In effect, the combined resistors total up to form one giant resistor. This can be expressed mathematically as:

$$R_T = R1 + R2 + R3, \text{ etc.}$$

Parallel Circuit

Another way resistors can be arranged is in a parallel formation. This, not surprisingly, is called a *parallel circuit*. In this type of circuit, the current divides into two or more paths before rejoining the circuit. Figure 2-15 shows three resistors combined in such an arrangement.

In a parallel circuit, total resistance is always less than that of the smallest resistor. The total circuit resistance can be determined by the following formulas:

$$R_T = \frac{R1 \times R2}{R1 + R2} \qquad \frac{1}{R_T} = \frac{1}{R1} + \frac{1}{R2} + \frac{1}{R3}$$

(two resistors)   (three or more resistors)

If there are more than three resistors, just keep repeating the second formula. Using these formulas to solve the problem in Fig. 2-15, the answer is 2.8 ohms.

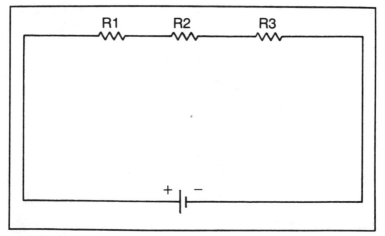

Fig. 2-14. Resistors in series.

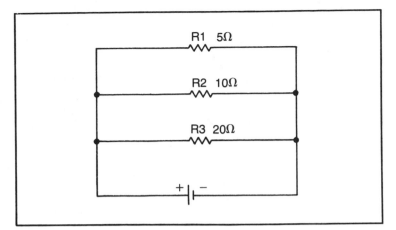

Fig. 2-15. Resistors in parallel.

Series-Parallel Circuit

Sometimes, we may have a circuit that is actually a combination of the series and parallel types. In a *series-parallel circuit* (such as shown in Fig. 2-16), you must determine the value of the resistors in the parallel circuit, then merely add that number to the resistance of the rest of the circuit. In other words, proceed as if the total resistance of the parallel circuit were just a large resistor in the series circuit.

PROJECT 3: CONDUCTANCE

As we've said, different materials, when placed in a circuit, offer varying amounts of resistance. Here's a way to see this statement in action. Merely take a small wooden board and insert two screws in it. Next, connect a battery and lamp to the screws as shown in Fig. 2-17. What we

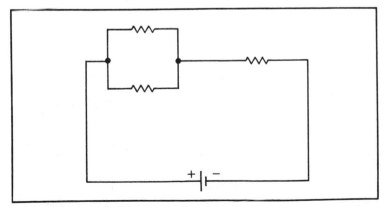

Fig. 2-16. Series-parallel circuit.

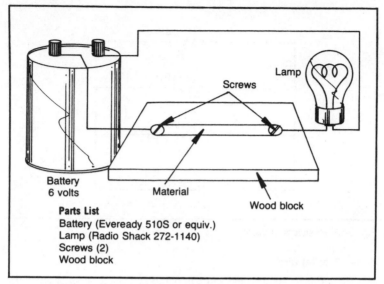

Fig. 2-17. Project 3: conductance.

have now is a circuit, minus a few inches between the screws. In this space, try inserting different materials—aluminum foil, rubber, wood—whatever you have around the house. You'll quickly notice that certain substances work better than others. We call this action *conductance*—the opposite of resistance. Depending upon the material used, the lamp will either shine

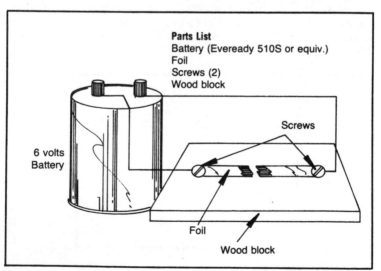

Fig. 2-18. Project 4: resistance.

brightly, dimly, or not at all. Silver, by the way, is a superb conductor—if you can afford it.

PROJECT 4: RESISTANCE

In Project 3 we saw how various materials conduct. Now, let's look at the opposite side of the coin—resistance. Using the same wooden block, connect it to the battery specified in the parts list shown in Fig. 2-18. Next, take some foil (the thin type found inside chewing gum or a cigarette pack) and place a *very narrow* strip of the material between the screws. Connect the battery and watch the foil burn away.

The foil burned because there was too much current for it to handle, causing it to overheat. If the strip were wider, thicker, or perhaps of a better conducting material, there would have been no problem. Instead, the electrons were forced to move through at such a rate that they disintegrated the foil.

Chapter 3

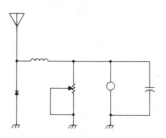

Magnetism and Induction

Earlier we saw how magnetism could help us produce alternating current. At that time, we intimated that magnetism plays a large role in the overall field of electronics. Now, let's take a closer look at magnetism and some of its other uses.

WHAT IS MAGNETISM?

Magnetism is an invisible force that exists around the immediate area of a magnet. Another name for this force is the *magnetic field*. Magnets themselves can be made from any number of different materials. A naturally occurring type of magnet is called a lodestone. In electronics, most magnets are made out of iron, steel, or ferrite.

Magnets have the unique property of being able to attract magnetic materials (objects capable of becoming magnets themselves). Most of us are acquainted with this phenomenon from the childhood game of picking up iron filings with a magnet (Fig. 13-1). From these childhood experiences, we also know that magnets have two opposite ends (called the *north* and *south poles*), and that opposite poles attract while like poles repel.

Like electrical flow, we cannot see magnetism, yet we can note its effects. One way we can observe the magnetic field's influence is by placing a bar magnet under a sheet of paper sprinkled with iron filings. Tap the paper once or twice and the pattern shown in Fig. 3-2 will result. The filings line up in a distinct pattern showing the magnetic force lines—a visual representation of the magnetic field.

TYPES OF MAGNETS

There are two different types of magnets—*temporary magnets* and

Fig. 3-1. A magnet picking up iron filings demonstrates magnetic attraction.

permanent magnets. As you may guess from their names, temporary magnets retain the bulk of their magnetic property for only a short time, while permanent magnets stay magnetized permanently.

An example of a temporary magnet would be soft iron, which is easily magnetized by placing it within the field of another magnet. A permanent magnet (which, by the way, is harder to magnetize), is usually made out of iron or steel.

ELECTROMAGNETISM

When current flows through a wire, it too creates a magnetic field. This *electromagnetism* can either be very incidental or significant, depending upon the configuration of the wire. To get the strongest magnetic field possible, wire is coiled up like a spring. Figure 3-3 shows a wire coil and its magnetic field.

Having a magnet we can turn on or off just by flowing current through a coil is very useful. *Relays* (such as found in many two-way radios) use this principle. If you own a video cassette recorder, you can hear relays (also called *solenoids*) clunking on and off as you switch from stop to play. But where have we seen wire coils used earlier in this book? That's right, in a generator.

Back in Chapter 1, if you'll remember, we had a wire coil spinning between two magnets to generate electric power. At the time, we men-

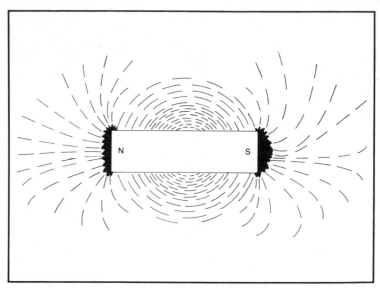

Fig. 3-2. Iron filings can give us a visual impression of the magnetic field.

tioned that whenever a wire loop spins through a magnetic field, voltage is induced within the wire. What we didn't explain back then was exactly how that happened. Now, let's take a closer look.

INDUCTANCE

Figure 3-4 shows a coil and a bar magnet. As you can see, there is no battery or other power generating device in the circuit. Yet, as you can also

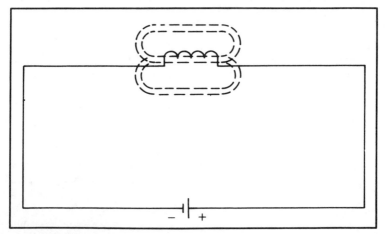

Fig. 3-3. The magnetic field created by a wire coil.

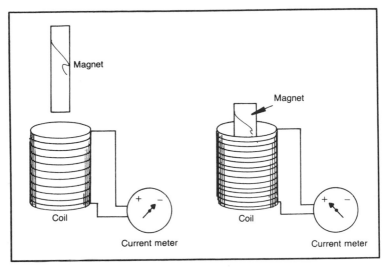

Fig. 3-4. A bar magnet inducing voltage in a coil.

see, when the magnet is pushed within the coil, voltage is generated and shows on the face of the meter. In a way, this sort of reverses the workings of a generator. Instead of the wire moving, the magnet moves. But the principle remains the same. As the magnetic field moves through the coil, voltage is induced within it. We call this induced voltage, *inductance*.

But is inductance used only for power generation purposes? No. Radio circuits, filters, power supplies for dc circuits, and dozens of other devices make use of this phenomenon.

Mutual Inductance

One of the most fascinating things a coil can do is transfer voltage from one circuit into another merely by coming in close proximity with another coil. We call this ability *mutual inductance*.

For instance, look at the circuit in Fig. 3-5. Side A has a coil connected to an ac power source. Side B has its coil hooked to a lamp. When the coils are moved near each other, the magnetic field from coil A induces a voltage in coil B, lighting the lamp.

The amount of effect between the two coils depends on a number of factors. Distance and angle between coils are the two most important considerations. Obviously, the closer the coils are to each other, the greater the effect. Likewise, the effect, or *coupling* as it is usually called, diminishes greatly when the coils are at right angles.

Inductors in a Circuit

When placed in a circuit, inductors behave in the same way as series or parallel connected resistors. This is to say that when hooked in series, the

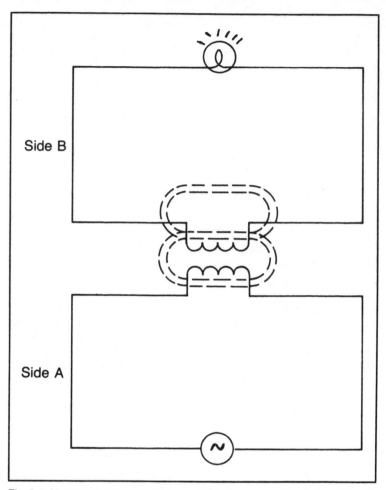

Fig. 3-5. Mutual inductance: coil in side A is inducing a voltage in side B's coil.

total inductor value is equal to the sum of the individual inductors. When placed in parallel, total inductance is always less than the value of the smallest inductor. These rules only apply, of course, as long as there is no coupling between inductors.

PROJECT 5: ELEMENTARY COMPASS

Speaking of magnetism, do you know that you live on a magnet? We all do! The earth is just one big magnet complete with a magnetic field, and North and South Poles. Don't believe it? Then try this experiment.

Take an empty bottle or jar and place within it, suspended from a thread, a needle punched through a thin piece of cardboard (Fig. 3-6).

Before inserting the needle in the bottle, however, rub it (in one direction only) on a pole of a magnet. Once in the bottle, the needle will always end up in a north-south direction, no matter which way you try to turn it. Proof that the earth is a magnet.

PROJECT 6: INDUCTANCE

As you know, a magnet moving through a coil induces voltage. Here's an easy way to see this action for yourself. As in Fig. 3-7, take some wire (just about any thin type will do) and circle it into a coil of about 25 or 30 turns. Now, run about two feet of wire from both ends of the coil over to an ordinary dime store compass. Wrap the wire about 25 or 30 times around the compass. Next, take a bar magnet and thrust it in and out of the coil as quickly as you can. Look at the compass. What do you see? We're taking the induced voltage through the wire and to the compass, making it deflect. Since the compass also has a coil around it, the magnetic field is exhibited on the compass' dial. Wouldn't it be great if we could power our homes this way? We'd get pretty sore arms, though!

PROJECT 7: ELECTROMAGNET

Playing around with electromagnets is a lot of fun, and they're very easy to make. Just take an ordinary pencil and wrap some wire around it. Then connect it to a cell, as shown in Fig. 3-8, and you're all set.

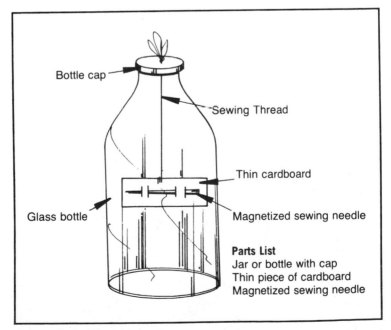

Bottle cap

Sewing Thread

Thin cardboard

Glass bottle

Magnetized sewing needle

Parts List
Jar or bottle with cap
Thin piece of cardboard
Magnetized sewing needle

Fig. 3-6. Project 5: a simple compass.

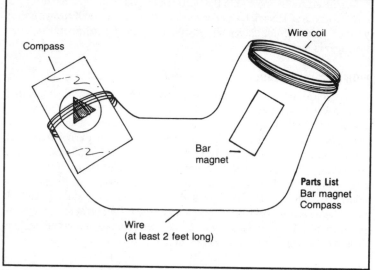

Fig. 3-7. Project 6: inductance.

Try varying the number of turns to see how it affects the magnet's power. You can also observe how electromagnets are used as relays. Switch the power on and off, and you'll see how quickly the magnetic field appears and disappears.

PROJECT 8: IMPROVED ELECTROMAGNET

The electromagnet in this project operates the same way as the one in

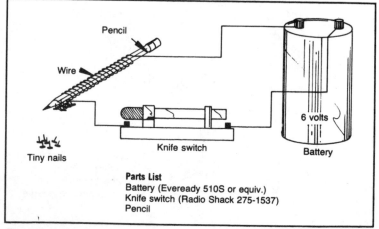

Fig. 3-8. Project 7: electromagnet.

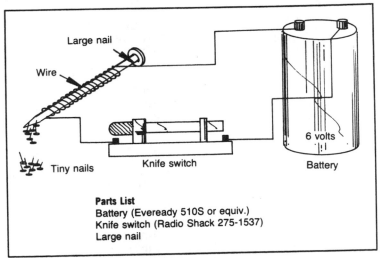

Parts List
Battery (Eveready 510S or equiv.)
Knife switch (Radio Shack 275-1537)
Large nail

Fig. 3-9. Project 8: improved electromagnet.

the previous example with only one important difference: the addition of a steel core. Our "core," in this case, is an ordinary nail.

Compare this electromagnet with the one in Project 7. See how many paper clips, tiny nails or tacks, etc., you can pick up with this unit, compared to one with the same number of turns without a steel core.

PROJECT 9: TELEGRAPH SYSTEM

The telegraph was the first electrical communications system developed by man. Interestingly enough, it operated through the application of electromagnetism.

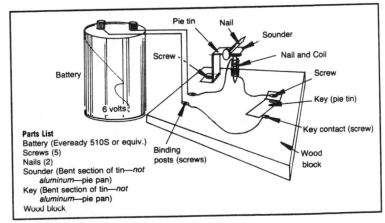

Parts List
Battery (Eveready 510S or equiv.)
Screws (5)
Nails (2)
Sounder (Bent section of tin—*not*
    aluminum—pie pan)
Key (Bent section of tin—*not*
    aluminum—pie pan)
Wood block

Fig. 3-10. Project 9: telegraph system.

41

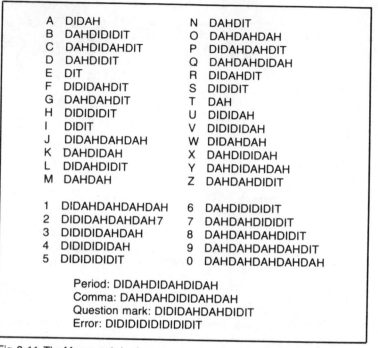

A	DIDAH	N	DAHDIT
B	DAHDIDIDIT	O	DAHDAHDAH
C	DAHDIDAHDIT	P	DIDAHDAHDIT
D	DAHDIDIT	Q	DAHDAHDIDAH
E	DIT	R	DIDAHDIT
F	DIDIDAHDIT	S	DIDIDIT
G	DAHDAHDIT	T	DAH
H	DIDIDIDIT	U	DIDIDAH
I	DIDIT	V	DIDIDIDAH
J	DIDAHDAHDAH	W	DIDAHDAH
K	DAHDIDAH	X	DAHDIDIDAH
L	DIDAHDIDIT	Y	DAHDIDAHDAH
M	DAHDAH	Z	DAHDAHDIDIT
1	DIDAHDAHDAHDAH	6	DAHDIDIDIDIT
2	DIDIDAHDAHDAH7	7	DAHDAHDIDIDIT
3	DIDIDIDAHDAH	8	DAHDAHDAHDIDIT
4	DIDIDIDIDAH	9	DAHDAHDAHDAHDIT
5	DIDIDIDIDIT	0	DAHDAHDAHDAHDAH

Period: DIDAHDIDAHDIDAH
Comma: DAHDAHDIDIDAHDAH
Question mark: DIDIDAHDAHDIDIT
Error: DIDIDIDIDIDIDIDIT

Fig. 3-11. The Morse code in phonetics. Try saying each character as you send it; that's the best way to learn.

Today, when we think about telegraphy, we usually associate it with the beeps coming out of a loudspeaker. But back in the early nineteenth century, before the speaker was even invented (not to mention the circuitry used to generate "beeps"), another way had to be found to communicate the sounds of the Morse code. That way was found through the use of a device called the *sounder.*

Figure 3-10 shows our version of a sounder, along with a key to tap out the code and a cell to provide power. When the key is depressed, a circuit is completed sending energy to the small electromagnet hammered into the system's wooden base. The magnetic field generated draws the bent piece of pie tin, and nail glued to it, to the electromagnet nail's head. This makes a click and gives us our Morse "sound." A simple invention, but a very important one. A phonetic representation of the Morse code is shown in Fig. 3-11.

Chapter 4

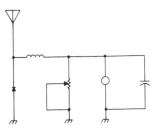

Power Supplies

Batteries do an excellent job of powering dc circuits, but they do have many obvious limitations. Ac generators are also wonderful sources of power, but you can't run alternating current through a dc circuit. Since there's no practical way of modifying a battery to deliver the power needed for all the different varieties of dc-operated devices, a way must be found to convert ac into dc. That's the job of the power supply.

POWER TRANSFORMERS

Before switching power from ac to dc, there's also the not-so-little matter of changing voltages. As you probably know, ordinary household ac runs at about 117 volts. But what if we have a device that needs, say, 12 volts? Pumping 117 volts into a 12-volt unit, be they ac or dc volts, will still result in a smoking lump of charcoal. A way must be found to *transform* high voltages to low.

The transformer (Fig. 4-1) is just the unit for this task. A rather formidable (some say ugly) looking device, the transformer is really quite simple in its construction. Operating on the theory of mutual inductance, a transformer's basic parts consist of two coils mounted in close proximity, some sheets of laminated steel to wind the coils around, and a housing to hold the entire thing together. Figure 4-2 shows a diagram of a typical power transformer.

In mutual inductance, as we've said, one coil induces a voltage in a neighboring coil. We've also noted that the amount of voltage induced in the nearby coil can vary, depending upon such factors as distance and angle. Since our idea is to use the transformer to reduce voltage to a usable level, we don't want a full voltage exchange between the two coils. The most

43

Fig. 4-1. The power transformer: it can either step-up or step-down ac voltage.

practical way to achieve this in a transformer is to differ the number of windings—*turns*—on each coil.

In a transformer, the first coil (connected to the ac source) is called the *primary*. The second coil is called the *secondary*. A transformer that reduces voltage from the primary to secondary, is called a *step-down transformer*. Similarly, there are also times when we need to *raise* the voltage from the primary to the secondary. Units that do this are called *step-up transformers*,

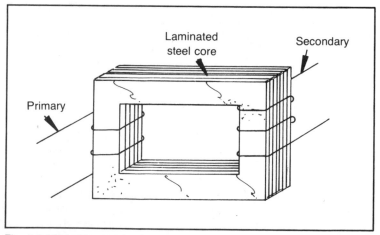

Fig. 4-2. Vital components of a power transformer.

and are used in circuits that require higher voltages than being delivered by the ac source.

There's a simple formula we can use to determine the relationship between volts and coil turns in any transformer. We call this the *voltage-turns ratio*, and it is expressed thusly:

$$\frac{E_P}{E_S} = \frac{T_P}{T_S}$$

where,

E_P = Primary voltage
E_S = Secondary voltage
T_P = Number of turns in primary
T_S = Number of turns in secondary

This means that in a step-up transformer with a primary voltage input of 117 volts and a secondary output of 351 volts, there are three times as many turns on the secondary coil as there are on the primary. Conversely, if we wanted to step-down the voltage, with three times as many turns on the primary, the secondary's output would be 39 volts.

Transferring Current

When we are transferring voltage between coils, we are also switching, of course, current. The formula for determining just how much current is being transferred is:

$$\frac{I_S}{I_P} = \frac{T_P}{T_S}$$

where

I_S = Secondary current
I_P = Primary current

This means that the ratio of primary current to secondary current varies in inverse proportion to the turns ratio.

Transformer Efficiency

One other factor can affect the relationship between input and output power: the efficiency of the transformer. Since we live in an all too imperfect world, no transformer will operate at 100% of its theoretical capability. Therefore, when designing transformers into a circuit, we must take note of its rated efficiency.

Let's say we have a step-down transformer that's made to operate at 80 percent efficiency. If we have an input of 117 volts with a power of 150 watts and a turns ratio of 20:1, how do we find the secondary power, secondary voltage, primary current, and secondary current? With these formulas:

$$P_S = P_P \times \text{Efficiency}$$ $$P_S = 150 \times 80; P_S = 120 \text{ watts}$$

$$E_S = \frac{T_S}{T_P} = E_P$$ $$E_S = \frac{117}{20} = 5.85 \text{ volts}$$

$$I_P = \frac{P_P}{E_P}$$ $$I_P = \frac{150}{117} \; ; I_P = 1.28 \text{ amperes}$$

$$I_S = \frac{P_S}{E_S}$$ $$I_S = \frac{120}{5.85} \; ; I_S = 20.5 \text{ amperes}$$

And that covers the basics of transformers.

RECTIFICATION

The process above leaves us with an ac voltage we are trying to change into dc. We do this via *rectification*. The word "rectification" means to rectify something—to correct it. And that, of course, is what we're going to do to ac: "correct" it into dc. After all we've gone through just to produce ac and alter its power, switching it over to dc is like child's play. Well . . . almost.

Rectifying ac into dc calls for an electronic component known as a *rectifier*. The rectifier most commonly used is called a *diode*. The diode has the unique ability of allowing current to flow in only one direction (Fig. 4-3). Using a sine wave to represent current, Fig. 4-4 shows what a diode does to ac. The negative halves of the cycle are lopped off, leaving us with voltage going only from a positive peak to zero. This is called *half-wave rectification*, since only half of the cycle is being used.

But is this dc? Well, yes and no. It's dc in that current is flowing in only one direction, but it isn't pure dc in respect to the rapid pulsing that occurs from the one-half of the cycle missing. At best, we could call it "pulsating dc." To make this dc more like the type we get from batteries, a way must be found to smooth the output voltage.

Full-Wave Rectification

One way we could smooth out the voltage is by finding a way to use the other half of the ac cycle. In order for us to have dc, of course, this second cycle half must move in the same direction as the first. We can do this by

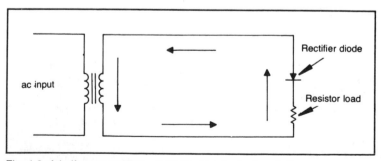

Fig. 4-3. A half-wave rectifier—it produces a "pulsating dc."

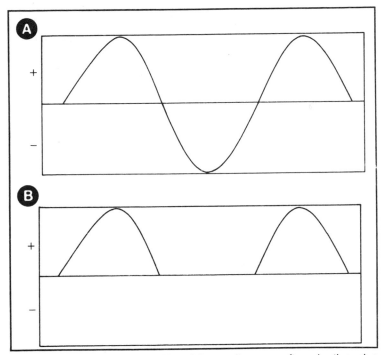

Fig. 4-4. A regular ac sine wave (A), and the way it appears after going through a rectifier diode (B).

adding another diode to the circuit and tapping into the power transformer at a point halfway through its secondary coil.

How does this give us two cycle halves with current flowing in the same direction? Easy. With two rectifiers, each taking outputs from opposite ends of the coil, we have double the work being done. The center tap effectively splits the secondary into two parts, and since the diodes will only pass current one way, the positive cycle peaks (Fig. 4-5) come one after another with no pause in between.

Full-Wave Bridge Rectification

We can also obtain full-wave rectification another, better way, by using yet two more diodes. We call this *full-wave bridge rectification*. The word "bridge" is included, because the four diodes are connected in an interlocking network. As the circuit in Fig. 4-6 shows, when side A of the transformer secondary is positive, diodes 1 and 3 conduct. When side B is positive, diodes 2 and 4 have current flowing through them. The main advantage here is that since we are using the entire secondary coil at all times, the bridge's output is double that of a standard full-wave rectifier; that's full secondary voltage at all times.

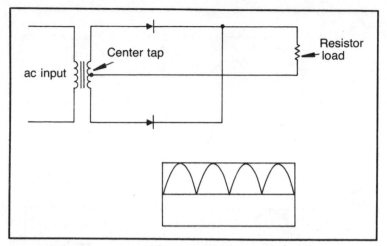

Fig. 4-5. A full-wave rectifier circuit and its resulting waveform. Positive peaks follow each other with no time interval. Compare this waveform to the half-wave type in Fig. 4-4.

Filtering

Full-wave rectification improves the "pulsing" situation quite a bit, but as you can see in Fig. 4-7, our voltage still isn't very steady. While the waiting time is eliminated, we still have peaks and depressions that must be smoothed.

One of the most popular methods of filtering voltage output is to use a *filter capacitor*. What this component does is to act like a reservoir—saving when there's an abundance, releasing when there's a shortage. When voltage is high, the capacitor stores; when voltage drops, the capacitor releases. This is expressed in Fig. 4-8. As you can see, by using the filter capacitor, the ripple is virtually eliminated, providing us with almost perfect dc—good enough for most applications.

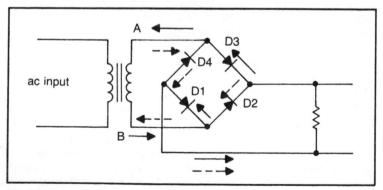

Fig. 4-6. Full-wave bridge rectifier.

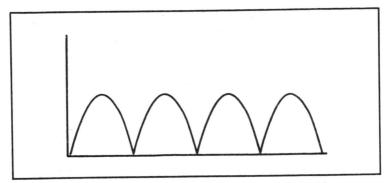

Fig. 4-7. Even after full-wave rectification, our dc voltage output still fluctuates up and down.

REGULATION

It's important that once we have our dc, and after it's been fully filtered, that we can depend upon an unvarying output being delivered to our circuit. Unfortunately, some power supplies have the nasty tendency of putting out less voltage as greater loads are applied. This can lead to many upsetting consequences. Imagine, for instance, a CB radio connected to such an undependable power supply. On receive, when demand is low, everything works fine. But once the mike button is depressed, the meter light dims, the digital display conks out, and the transmitter emits only a weak, warbly signal.

We prevent such sad happenings by using a *regulated* power supply. Regulation is achieved via a number of methods—zener diodes, transistors, and integrated circuits are all used to hold voltage at a steady level. We'll see all three of these component types used in projects at the end of this chapter. For now, remember that voltage regulation is an important consideration when building or buying power supplies.

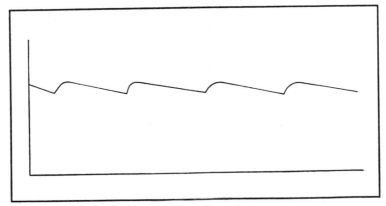

Fig. 4-8. With filtering, dc voltage is almost perfect.

VOLTAGE MULTIPLICATION

We know the step-up transformers can raise voltage from primary to secondary coils. It's also possible, however, to increase voltage after it leaves the transformer. We call circuits that do this *voltage multipliers*. Figure 4-9 shows such an arrangement at work. In this example, voltage is being doubled, so we call the circuit a *voltage doubler*. Other multiplier circuits can be made to triple or even quadruple voltage output.

As the example shows, a voltage multiplier is basically a half-wave rectifier employing a pair of diodes and capacitors. With 117 volts coming out of the transformer secondary, when side A is negatively polarized—side B positive—diode 1 conducts and capacitor 1 charges to 117 volts. When the polarity reverses, side A positive, side B negative; then diode 2 conducts and capacitor 2 charges up to 234 volts. Why 234 volts? Because capacitor 2 is in series with capacitor 1, so it receives voltage from both the first capacitor and the ac line. If this doesn't make total sense, examine the diagram carefully. Remember, diodes can only pass voltage in one direction and capacitors alternately store and release voltage.

Since multipliers are half-wave rectifiers, they produce badly rippled dc—and their regulation isn't too good either. Lots of filtering can help smooth the output, but because of poor regulation, these circuits are mostly used in places like television power supplies—applications where high voltage is necessary, but demand fairly constant.

In short, multiplication can best be termed a "cheap and dirty" way of boosting voltage. Its main benefit comes from the fact that high voltage power transformers are very expensive, so voltage multipliers are an attractive way of cutting costs.

PROJECT 10 : 12 VOLT REGULATED POWER SUPPLY

Power supplies: should you build or buy? It's not an easy choice, but let us put in a few words for "brewing your own." Pre-built power supplies (Fig. 4-10) are available at most retail electronics stores for a fair price, but a home-built supply has many inherent advantages. Besides the fact you'll be able to custom-design the unit to your own specifications, a home-built

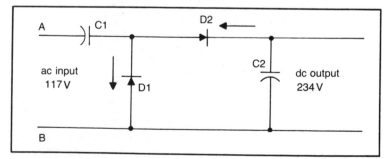

Fig. 4-9. Voltage doubler—a cheap and dirty way of boosting voltage.

Fig. 4-10. A commercially manufactured power supply.

supply affords you the opportunity to save a few bucks. Many parts can be salvaged from old appliances, and even if you must buy new components, at least you'll know you're getting quality workmanship for your money—your own! Anyway, building is fun.

Figure 4-11 shows a schematic of a power supply you'll be able to put together in only a few hours. From looking at the circuit, we can see that this supply uses a full-wave bridge rectifier to produce a dc output of about 12 volts. Regulation is achieved through the use of a zener diode and a transistor.

Although this is the most complicated project presented in this book so far, construction is really very easy. Just connect the components together on a small breadboard, making sure that your solder connections are clean and bright. One very important word of warning: *household ac power can kill! Be careful!* All sections of this circuit before the transformer primary have full line voltage running through them, so *never* touch any component in this area while the unit is plugged in. For added safety, this power supply should be mounted in a small metal cabinet so dangerous voltages are never accidentally encountered.

Built neatly and safely, this power supply has literally hundreds of uses—as a calculator or pocket radio battery eliminator, for instance—built carelessly, it can kill you, a friend, or a loved one. We want you to live so you can read the rest of this book; safety first, last, and always!

PROJECT 11: 15-VOLT IC REGULATED POWER SUPPLY

More and more, integrated circuit regulators are making their presence felt in the power supply field. A majority of today's commercially-made supplies use ICs as regulators because of their high reliability and cheap-

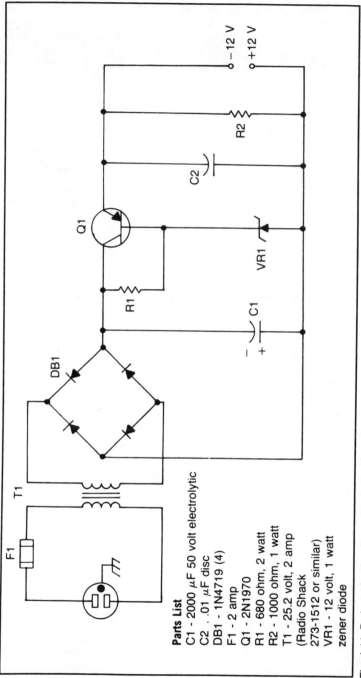

Parts List
C1 - 2000 μF 50 volt electrolytic
C2 - .01 μF disc
DB1 - 1N4719 (4)
F1 - 2 amp
Q1 - 2N1970
R1 - 680 ohm, 2 watt
R2 - 1000 ohm, 1 watt
T1 - 25.2 volt, 2 amp
(Radio Shack
273-1512 or similar)
VR1 - 12 volt, 1 watt
zener diode

Fig. 4-11. Project 10: 12 volt regulated power supply.

ness. While we'll be looking at ICs more closely in a later chapter, here's a power supply you can build using these devices right now (Fig. 4-12).

The most unusual aspect of this power supply is that we'll be converting a 25 volt single secondary transformer into a unit with two completely independent secondaries, each delivering 12.5 volts. To accomplish this, find the center-tap lead and follow it through the transformer's outer paper covering until you discover the point where it's connected to the secondary coil—use a sharp hobby knife for the cutting. Next, gently disconnect the two wires joined to the tap and solder in a new pair of outputs. Wrap some electrical tape around these leads so they won't short together and *presto!*—our old single secondary transformer has cloned itself a partner.

The remainder of the construction is straightforward, and shouldn't take more than a few hours. Be sure to neatly mount the unit in a metal case and ground connections to the chassis where indicated in the schematic.

PROJECT 12: NICKEL-CADMIUM BATTERY CHARGER

Rechargeable nickel-cadmium batteries are certainly a great money-saver over the course of their lifetimes, but you do need a charger to bring them back to life. You can either buy this unit at a store, or you can save yourself some money and whip one together from the circuit in Fig. 4-13.

Briefly, this circuit will allow you to charge up to 15 nicad cells at a time. The transistor and diode combination automatically limit the amount of current being delivered to the cells in proportion to the number of nicads being charged. Like the two projects before this one, be sure to mount your charger in a metal cabinet for safety. If you wish, you may use a standard battery holder to secure your nicads while they are being charged. This unit can be mounted on the outside of the charger, and may be swapped with holders designed for different cell sizes. Depending on the nicad type, full charge should be achieved in about 10-15 hours.

PROJECT 13: OVERVOLTAGE PROTECTOR

While most power supplies provide a reliable voltage output, occasional problems may develop from a defective regulator causing harmful voltages to be emitted. Likewise, poor battery connections may produce periods when voltage exceeds normal maximum levels. Whatever the cause, unwanted high voltage can wreck havoc on electronic gear and cost the user many dollars in repairs. To help prevent this problem, you may be interested in adding the circuit described here to either your power supply output or power input of your dc operated device.

Figure 4-14 shows a schematic of a very simple overvoltage protector. While many hobbyists rely solely on fuses to protect their equipment, these components take a relatively long time to work. By the time a fuse blows, irreparable harm may have already occurred. Our overvoltage protector, on the other hand, reacts immediately to control or shut down voltage before damage can take place.

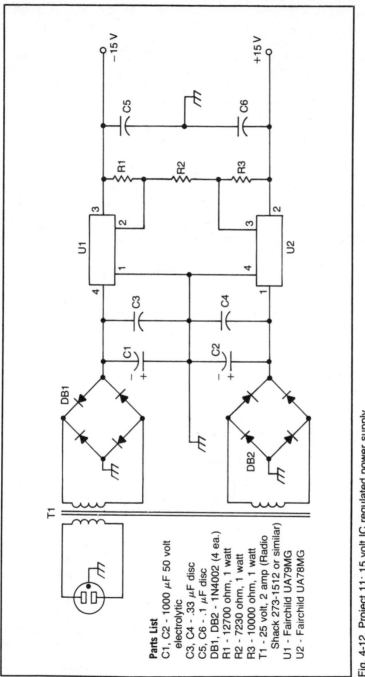

Parts List

C1, C2 - 1000 μF 50 volt
  electrolytic
C3, C4 - .33 μF disc
C5, C6 - .1 μF disc
DB1, DB2 - 1N4002 (4 ea.)
R1 - 12700 ohm, 1 watt
R2 - 7230 ohm, 1 watt
R3 - 10000 ohm, 1 watt
T1 - 25 volt, 2 amp (Radio
  Shack 273-1512 or similar)
U1 - Fairchild UA79MG
U2 - Fairchild UA78MG

Fig. 4-12. Project 11: 15 volt IC regulated power supply.

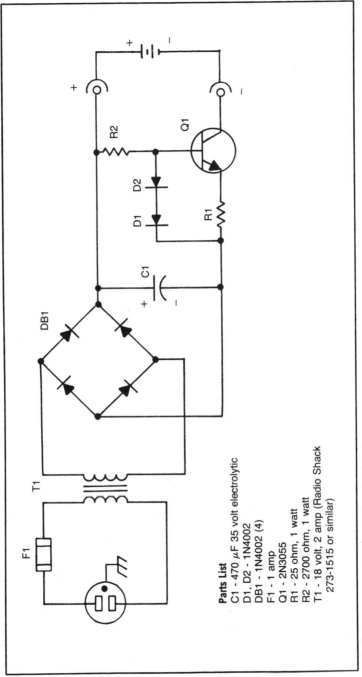

Parts List
C1 - 470 μF 35 volt electrolytic
D1, D2 - 1N4002
DB1 - 1N4002 (4)
F1 - 1 amp
Q1 - 2N3055
R1 - 25 ohm, 1 watt
R2 - 2700 ohm, 1 watt
T1 - 18 volt, 2 amp (Radio Shack
    273-1515 or similar)

Fig. 4-13. Project 12: nickel-cadmium battery charger.

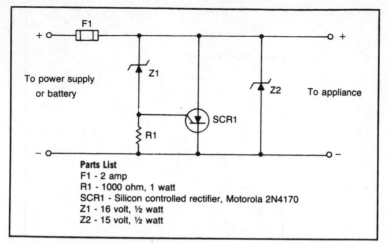

Parts List
F1 - 2 amp
R1 - 1000 ohm, 1 watt
SCR1 - Silicon controlled rectifier, Motorola 2N4170
Z1 - 16 volt, ½ watt
Z2 - 15 volt, ½ watt

Fig. 4-14. Project 13: overvoltage protector.

In this circuit, when voltage unexpectedly rises, Z2 acts to hold output at a safe level for a limited period of time (in case the voltage surge is just transient). But should the unusually high voltage continue, Z2 will burn out and cause Z1 to fire the SCR. The SCR then shorts the input voltage and blows the fuse. The user is now free to investigate the problem without having sacrificed his equipment. A new zener diode and fuse, and the overvoltage circuit is ready to work again.

Note: this circuit works well as-is with Project 10. For use with Project 11, use zeners rated three volts higher; the safety level will then also be raised three volts. In its original configuration, this circuit will also give protection to devices operated from automobile power supplies.

Chapter 5

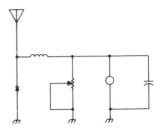

Capacitance and Capacitors

So far, we've seen two factors that can affect current flow: resistance and inductance. A third way we can affect current is through an action known as *capacitance*. Measured in *farads* (or, more commonly, in smaller units of microfarads and picofarads—μF, pF), capacitance is the property of two conductors to store an electrical charge. The device we use to implement this phenomenon is called a *capacitor*.

Like many of the other components we've examined, capacitors come in many forms. We have *electrolytic capacitors* (Figs. 5-1 and 5-2), *metalized film capacitors* (Fig. 5-3), *tantalum capacitors* (Fig. 5-4), *trimmer capacitors, disc capacitors*, and dozens of other variations. Each capacitor type is selected by circuit designers for various individual properties and characteristics, but they all have one important fact in common: they store electrical charge. If they didn't, they wouldn't be capacitors.

CAPACITOR THEORY

You'll remember from the preceding chapter that we used a capacitor as a filter to smooth out the ripple in dc voltages. This was done by having the capacitor save energy when it flowed through the circuit, then releasing it when voltage fell off. But exactly how does a capacitor manage to do this little trick? How is it able, in effect, to switch itself on and off?

If we were to cut open a capacitor, any capacitor, we would discover that it is made up of two metal *plates* separated by a *dielectric*—a material that can't conduct electricity (Fig. 5-5). When a capacitor is placed in a circuit (such as the one shown in Fig. 5-6), electrons flow from the negative terminal of the battery to plate A of the capacitor. They do this because electrons, as we know, always move to places with a lesser negative

Fig. 5-1. Various sizes of electrolytics. The ink bottle is included for size comparison. This is the type of capacitor you might find used as a power supply filter.

charge. Since the battery's negative terminal has an abundance of electrons, they quickly rush to the capacitor plate. Plate B, for its part, is losing electrons to the battery's positive terminal at the same rate as electrons are arriving at A, rendering B positively charged.

We can now say that plate A is negatively charged in relation to plate B. Normally, the electrons on A should move over to B in their never-ending attempt for balance. However, separating the two plates is the dielectric,

Fig. 5-2. Some more electrolytics. These are polarized, meaning they can be charged in one direction only. Other electrolytics are non-polarized. Polarized electrolytics, obviously, can't be used in ac circuits.

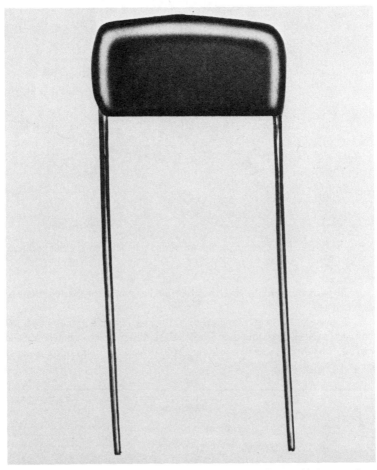

Fig. 5-3. Metallized film capacitors, like this one, work well over wide temperature ranges.

and since it's unable to conduct, the electrons can't move. If we were to take a small piece of wire made from a good conductor (like copper) and place it over the capacitor's leads, circumventing the dielectric, we would see a spark as the electrons are suddenly released from plate A and flow across the conductor to B.

Capacitor Charging

Besides insulating the two plates, the dielectric also works to control how many electrons move from the battery to plate A. It's able to do this because of an effect passed through it by the charge of plate B. In a way, this effect—called the *electric field*—is much like the magnetic field we ob-

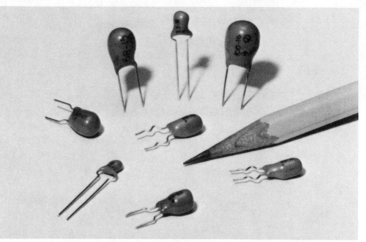

Fig. 5-4. Tantalum capacitors pack a lot of capacitance into a small package.

served earlier. The attraction (generated by the positive charge of B) actually works to hold the electrons on A and assures that they continue to accumulate.

The electrons keep piling up until they achieve a force equal to the voltage of the battery. In other words, if we have a six volt battery in the circuit, the capacitor will store six volts. We call this entire process *capacitor charging*.

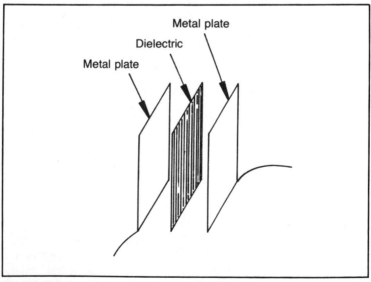

Fig. 5-5. The three basic elements of a capacitor.

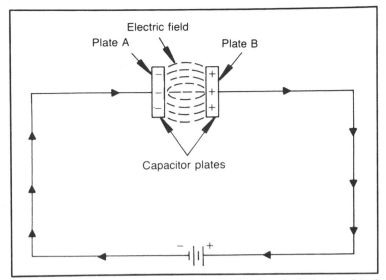

Fig. 5-6. The capacitor in a dc circuit: electrons flow from the battery to plate A and also out of plate B to the battery. Electrons are then held on plate A by the electric field — a mutual effect between the two plates.

Capacitor Discharging

If we desire, we can carefully remove the capacitor from our circuit and place it in another circuit containing a lamp (Fig. 5-7). As soon as we do this, the capacitor quickly unloads its charge and lights the lamp for a brief instant. The reason behind this action is quite simple. The lamp's filament connects the plates together. The electrons on plate A, anxious to move to positively charged plate B, rush through the lamp, light it, and wind up on the second plate. The capacitor then has an equal number of electrons on each plate and is ready to be charged again. Actually, we just used the lamp as an example. Any object placed in the circuit to introduce resistance would do approximately the same job. This process is called *capacitor discharging*.

Capacitor Characteristics

The total capacity of a specific capacitor depends upon a number of factors. The area of the plates, the distance between them, and, perhaps most important, the dielectric material. Since various insulators are affected differently by the electric field, capacitor storage can be greatly altered by changing dielectrics. For instance, glass dielectrics have a capacitance roughly two times better than mica, and mica works six times better than air. This is known as the *dielectric constant*, and is an important consideration for designers who must know just how much capacitance to introduce into a circuit using a given voltage, in a given application.

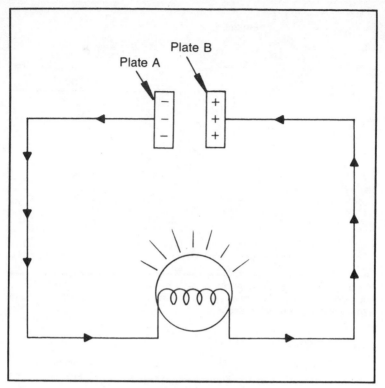

Fig. 5-7. As the capacitor discharges, it lights the lamp.

Another important consideration when selecting a capacitor is just how much voltage it can take before breaking down. All capacitors are designed to withstand just as much voltage—exceed the maximum capability and you'll ruin the capacitor. So if you have a capacitor rated for 250 WVDC (*working volts dc*) the component should not be used in a circuit exceeding this voltage. Unless you like fires and explosions, that is.

CAPACITORS IN DC CIRCUITS

Just like resistors and inductors, there are formulas used for determining the total capacitance of capacitors in a circuit. Perhaps the easiest way to learn how capacitors are added in series and parallel, is to remember that they react exactly opposite to resistors and inductors in the same type circuits. This is to say that while resistors and inductors in series are the sum of their individual component values, capacitors have a series capacitance always less than that of the lowest-value capacitor. In parallel circuits, the same is true. Capacitors in parallel are the sum of their individual capacitances, while resistors and inductors have a total value that's less than their smallest component. Just to be sure, here are the formulas:

$$C_T = \frac{C1 \times C2}{C1 + C2} \qquad \frac{1}{C_T} = \frac{1}{C1} + \frac{1}{C2} + \frac{1}{C3}$$

(two capacitors)   (three or more capacitors)   Capacitors in series

$$C_T = C1 + C2 + C3 \text{ etc.}$$

Capacitors in parallel

CAPACITORS IN AC CIRCUITS

To this point, we've concentrated on how capacitors behave in dc situations. However, it's also possible to use these components in ac circuits; and, as you'll soon find out, capacitors in an ac environment are totally different animals.

Since current in an ac circuit flows in two directions, a capacitor is subject to electrons striking each of its plates many times in a given second. When the current polarity changes, so does the flow of electrons striking the capacitor's plates. Fig. 5-8 shows this in detail.

As Fig. 5-8 also shows, current never actually passes through the capacitor, it just sort of bounces back and forth through the circuit. When the cycle is positive, electrons flow out of the right side of the generator and accumulate on capacitor plate B. When the cycle is negative, electrons rush to plate A. Since electrons are also flowing from the opposite plate in each instance, the result is a circuit that acts if voltage were actually passing through the capacitor.

To illustrate this concept better, let's say we place a lamp in a dc circuit containing a capacitor (Fig. 5-9). What happens? Nothing, of course. The capacitor is storing energy and won't let any of it get through to the lamp. But, on the other hand, let's power this circuit with an ac generator (Fig. 5-10). Now the lamp is lighting. Why? Because the power is flowing in and out of the capacitor's plates, acting for all the world as if it were actually passing through the capacitor.

What conclusions can we draw from this? Well, as a rule of thumb, we can say that capacitors in a dc circuit block the movement of electrons; in an ac circuit, they allow electrons to flow.

REACTANCE

Of course, capacitors do have an affect on an ac circuit. After all, if capacitors only worked to pass ac unhindered, there wouldn't be any reason to put them in the circuit in the first place. What capacitors do add in an ac circuit is an influence called *capacitive reactance.*

Perhaps the best way to describe reactance is to say that it is the opposition presented to alternating current by capacitance—sort of like resistance in a dc circuit. As a matter of fact, we even measure reactance in ohms, just like resistance.

The amount of reactance in a circuit depends upon two factors: the value of the capacitor and the frequency of the generator. There's a rela-

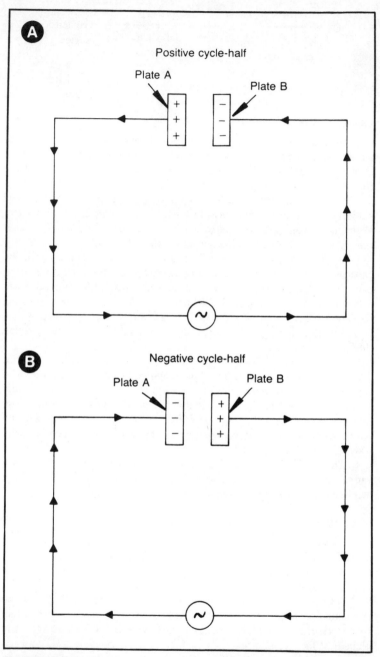

Fig. 5-8. In an ac circuit, current flows in two directions, alternately charging capacitor plates with opposite polarities many times in a second.

Fig. 5-9. Place a capacitor, in series, in a dc circuit, and the lamp doesn't light.

tively easy formula we can use to determine a capacitor's reactance in an ac circuit:

$$X_c = \frac{1}{2\pi fC}$$

X_c = capacitance reactance in ohms
π = 3.14
f = ac frequency
C = capacitance in farads

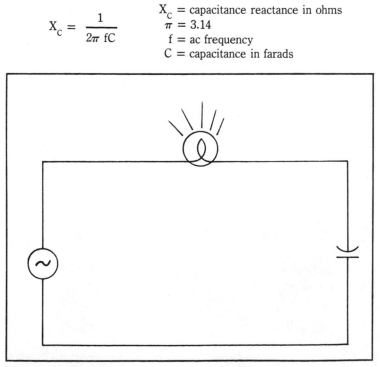

Fig. 5-10. Place a capacitor, in series, in an ac circuit, and the lamp lights.

Capacitors aren't the only components that provide reactance; inductors can also do this. We call this *inductive reactance* (also measured in ohms) and express its formula thusly:

$$X_L = 2\pi \, fL$$

X_L = inductive reactance in ohms
π = 3.14
f = ac frequency
L = coil inductance measured in "henrys"

Why are inductive and capacitance reactances important? Well, like most other principles we've discussed in this book, reactance can be a tool. Coils are often used to oppose or "choke" flows of alternating current to keep them from places they aren't wanted. Likewise, capacitors also have their uses in situations where we want to oppose the flow of alternating current. Some places where we use capacitors and inductors for reactance are in television interference filters (Fig. 5-11), power line filters (Fig. 5-12), and tuners designed to match radio transmitters to their antennas (Fig. 5-13). We'll be looking at these circuits in greater detail later on.

MEASURING CAPACITANCE

Getting back to capacitor measurement for a moment, as we said, a capacitor is measured in farads and its subdivisions. We say a capacitor is valued at one farad if it accepts one *coulomb* of charge when hooked to a one volt source. The coulomb is called the unit of quantity, and is equal to 6.28×10^{18} electrons.

CAPACITOR SAFETY

Capacitor safety? In a way, it's strange to think that a single electronic component can harm us, but it's true. After a circuit is disconnected from its power source, resistors and inductors lose all their energy, but capacitors can store their charge for days, even weeks, before they unload their store of electrons.

For example, let's say we have a circuit with a capacitor storing 600

Fig. 5-11. This television interference filter uses an inductor capacitor (LC) circuit to reduce the interference generated by CB radios.

Fig. 5-12. An LC circuit is also employed in this power line filter to reduce interference to computers and other sensitive equipment.

volts. After power is removed, energy is lost throughout the circuit—everywhere except in the capacitor, that is. A few hours later, an unwitting hobbyist comes along and sticks his fingers across the capacitor's leads. The next thing you know, he's on his way to his local hospital or, even worse, his neighborhood mortuary. All from a circuit that was disconnected hours ago.

Still, a way must be found to work on our circuit. After all, repairs must be made, and we just can't wait around until all the capacitor's energy slowly

Fig. 5-13. You can imagine the sizes of the inductor and capacitor in this antenna-transmitter matching device.

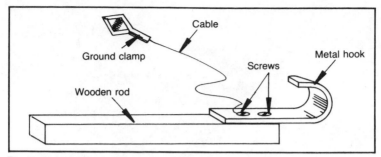

Fig. 5-14. Use this device to short high voltage capacitors.

leaks away. We remove charge from capacitors by using devices called *shorting sticks.*

A shorting stick does exactly what its name implies—it shorts the charge out of a capacitor and harmlessly into the ground. You can either make one of these simple tools yourself (from the illustration in Fig. 5-14), or you can purchase one at your local electronics or electrical supply store. A shorting stick is something no serious electronics experimenter should be without. Thousands of dead hobbyists could testify to their desirability, if only they could speak.

PROJECT 14: LAMP FLASHER

Back in the days before solid-state components made their presence felt, capacitors were often used to control neon lamps in various types of flashing circuits. These flashing lamps might be used in advertising signs, children's toys, or a host of other applications. While these circuits aren't used much today, they do make good examples of just how capacitors store and release charge. Practically speaking, you can easily incorporate these circuits into devices of your own—they make great science fair projects for kids.

In this project, we'll be flashing a single neon lamp. As you can see from the schematic in Fig. 5-15, a fairly high voltage battery is needed to power the lamp. If possible, try to secure the use of a test bench dc power supply. If you can't, then try linking together, in series, the batteries mentioned in the parts list.

In operation, the circuit works this way: current is drawn by C1 from the power source through R1. R1 acts to slow the flow of current so C1 can charge slowly, giving us a steady flashing cycle. When the capacitor is fully charged, the neon lamp begins conducting and soon drains the capacitor. The neon lamp then goes dark, and the capacitor begins to charge again. All in all, a classic example of capacitor charge and discharge.

PROJECT 15: TWO-WAY LAMP FLASHER

Can a capacitor control two neon lamps in a flasher circuit? You bet! We call this sort of circuit a "seesaw" for a pretty obvious reason. When one of the lamps is lit, the other is dark, and vice versa—they alternate, just like a seesaw.

In this circuit (Fig. 5-16), our capacitor is being charged from two different directions. When NE1 is lit, the capacitor charges through R2. When NE2 lights, the capacitor receives voltage through R1. The reason for this is very elementary. As the capacitor approaches full charge, the current flow slackens, causing NE1 to wink out. At that point, charging begins in the opposite direction and lights NE2. The alternating flashing will continue until power is removed.

PROJECT 16: THREE-WAY LAMP FLASHER

Would you believe three lamps? Yep! In this project (Fig. 5-17), we use three lamps, three resistors, and three capacitors to form a circuit arrangement that causes the neon lamps to light in endless succession—around and around and around.

Please note that this circuit uses about 60 more volts than the previous two, and that the neon lamps should be of an identical value and manufacture. It's very important that each of the lamps will conduct at the same voltage, otherwise the project won't work.

PROJECT 17: LIGHT-CONTROLLED SWITCH

Here's an intriguing little circuit (Fig. 5-18): a light-controlled switch

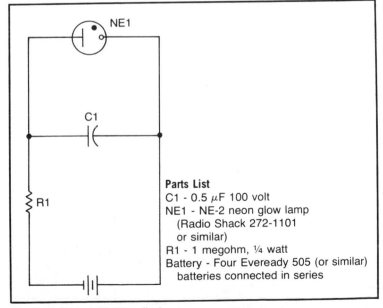

Parts List
C1 - 0.5 μF 100 volt
NE1 - NE-2 neon glow lamp
  (Radio Shack 272-1101
  or similar)
R1 - 1 megohm, ¼ watt
Battery - Four Eveready 505 (or similar)
  batteries connected in series

Fig. 5-15. Project 14: lamp flasher.

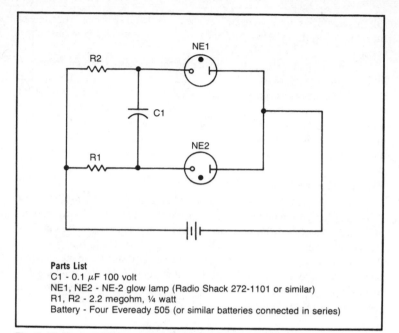

Parts List
C1 - 0.1 μF 100 volt
NE1, NE2 - NE-2 glow lamp (Radio Shack 272-1101 or similar)
R1, R2 - 2.2 megohm, ¼ watt
Battery - Four Eveready 505 (or similar batteries connected in series)

Fig. 5-16. Project 15: two-way lamp flasher.

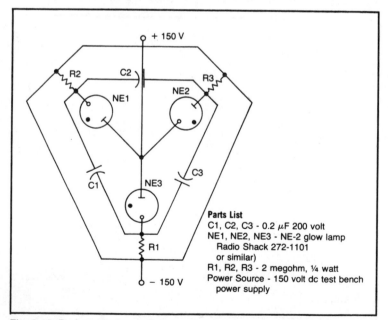

Parts List
C1, C2, C3 - 0.2 μF 200 volt
NE1, NE2, NE3 - NE-2 glow lamp
   Radio Shack 272-1101
   or similar)
R1, R2, R3 - 2 megohm, ¼ watt
Power Source - 150 volt dc test bench
   power supply

Fig. 5-17. Project 16: three-way lamp flasher.

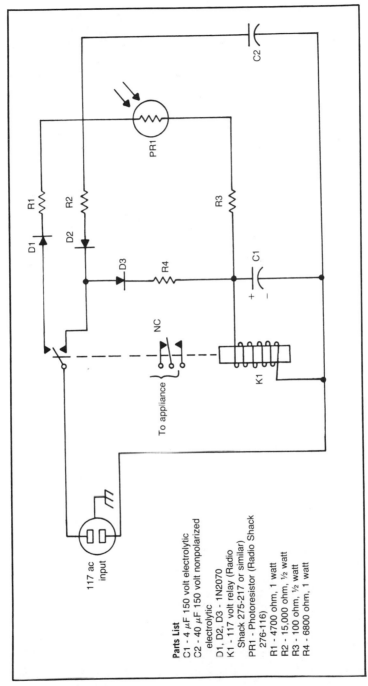

Parts List

C1 - 4 μF 150 volt electrolytic
C2 - 40 μF 150 volt nonpolarized
  electrolytic
D1, D2, D3 - 1N2070
K1 - 117 volt relay (Radio
  Shack 275-217 or similar)
PR1 - Photoresistor (Radio Shack
  276-116)
R1 - 4700 ohm, 1 watt
R2 - 15,000 ohm, ½ watt
R3 - 100 ohm, ½ watt
R4 - 6800 ohm, 1 watt

Fig. 5-18. Project 17: light-controlled switch.

that can be used to control an electric garage door opener, as a TV or radio remote control, or just about any other application you can think of. Whenever a beam of light strikes photoresistor PR1, the relay will close; flash some light again and the relay opens. This unit can either be mounted in a small metal cabinet of its own or, if you're ingenious, in the case of the appliance to be controlled. In any event, be sure that the photoresistor is shielded from ambient light. You can do this by placing PR1 at the end of a light-tight paper tube so only a head-on flashlight or automobile headlight beam can access it.

At the heart of this gadget is a nonpolarized electrolytic capacitor (C2). We use an electrolytic for a very good reason: since this is a fairly high voltage circuit, an electrolytic is the best capacitor choice for the job. The capacitor must be nonpolarized, though, since electrons will be striking it from two directions.

When power is applied, C2 is charged through D1 and R1. As light is shined on PR1, its resistance lowers and the charge flows off of the capacitor through R3 and the relay. Once the relay closes, C2 charges from D2 and R2, the flow coming from the opposite direction than before. When the light once again strikes the photoresistor, C2 discharges in reverse and opens the relay.

Chapter 6

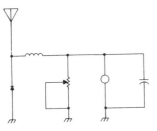

Series- and Parallel-Resonant Circuits

"Series-resonant circuit." Sounds ominous, doesn't it? Unfortunately, many electronics terms sound very scary when you first encounter them, but once you get acquainted with what the words actually stand for, the fear soon dissipates. This term is a classic example, so let's analyze the words to see exactly what they mean.

The word "series" is simple enough. We know what a series circuit is: a combination of elements in direct succession, as opposed to a parallel circuit where the components are hooked across each other. The next word is "resonant." Looking in our handy dictionary, we see it's derived from "resonance," which means to produce a large signal from a small stimulus. Our final word is "circuit," and if you don't know what that means by now, we kindly suggest that you reread this book from the beginning. What we end up with is a series circuit that produces an increased signal. That, in essence, describes a series-resonant circuit.

SERIES-RESONANT CIRCUIT ACTION

Look at Fig. 6-1. Here we have a variable inductor, a fixed capacitor, a lamp, and a household ac power source of 117 volts all hooked in series configuration. Watch what happens as we increase inductance by moving an iron core through the coil. The deeper it goes in, the more brightly the lamp shines. Suddenly, however, the core passes a point where the lamp starts to grow dimmer. Why? Because the greatest amount of current flows when induction reaches a critical point in relation to the capacitor. We call this point *resonance*, and since the components are connected in series, we call the entire shooting match a *series-resonant circuit*.

Sounds good. But why is it "resonant?" Well, think of a trumpet. Can

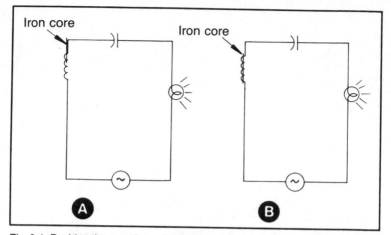

Fig. 6-1. Pushing the core through the inductor, current rises until it hits the point of resonance, where it flows at a maximum (A), then it diminishes (B).

you imagine how it would sound without its bell? Pretty awful. The bell takes the wispy noise from the instrument barrel and builds the sound up to a full, rich level. The same is true for a resonant circuit. When the circuit is out of resonance, current is weak. Put it into resonance and the current flow becomes very strong.

Getting back to our circuit, if we analyze it closely we discover that resonance occurs at a point where capacitive and inductive reactances are equal. As we remember from the previous chapter, reactance opposes the flow of ac. But when the two reactances reach an equal level (Fig. 6-2), they effectively cancel each other out and let ac flow unhindered.

Resonance is greatly affected by the frequency of a circuit. If the ac frequency is raised, the inductive reactance will rise while capacitive reactance falls. At this point, the current flow will decrease and we say that the circuit is no longer in resonance. If the frequency is lowered, an opposite action takes place, with capacitive reactance rising and inductive reactance dropping.

This probably won't surprise you, but we have a formula we can use to demonstrate that either inductance or capacitance can be changed to alter resonance. It is:

$$f_0 = \frac{1}{2\pi\sqrt{LC}}$$

f_0 = Resonant frequency (in hertz)
π = 3.14
L = Inductance (in henrys)
C = Capacitance (in farads)

Or, in more common values:

$$f_0 = \frac{10^6}{2\pi\sqrt{LC}}$$

f_0 = Resonant frequency (in hertz)
π = 3.14
L = Inductance (in microhenrys)
C = Capacitance (in microfarads)

Quite a couple of formulas. If you can solve them, great. If not, don't sweat. This is something circuit designers have to worry about, and since you're just getting your feet wet in electronics, we don't want to overburden you. We're including these formulas just to show you a way does exist to find resonant frequency by changing inductive and capacitive values.

PARALLEL-RESONANT CIRCUITS

Here's the other side of the issue—*parallel-resonant circuits*. Again, by analyzing the circuit's name, we can deduce that it is a resonant circuit in parallel configuration. The trick to this circuit, however, is that it reacts in an exactly opposite way to series-resonant circuits.

Looking at the parallel-resonant circuit in Fig. 6-3, using the same component values as in our previous circuit, we see that when the iron core is moved through the inductor, the lamp shines progressively dimmer. Then, as we pass the point of resonance, it begins glowing brighter. This tells us that in a parallel-resonant circuit, at resonance, current falls to a minimum.

Like its series brother, parallel-resonant circuits fall out of resonance if you change their frequency. To bring the circuit back into a resonant

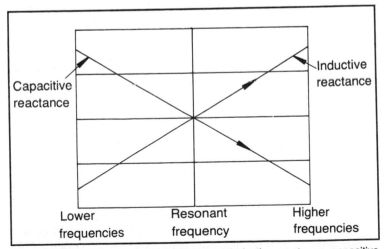

Fig. 6-2. Plotting the relationships between inductive reactance, capacitive reactance, and resonant frequency.

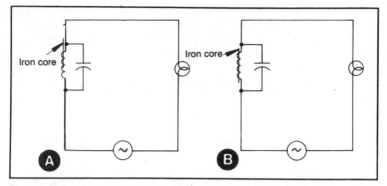

Fig. 6-3. Resonance in a parallel circuit: current falls until it reaches the point of resonance, where it flows at a minimum (A), then it starts rising again (B).

condition, you would have to change the values of either the inductor, the capacitor, or both. To find the frequency at which the component values would cause resonance, we would use the same formula as applied to series-resonant circuits.

RESONANT CIRCUIT USE

Resonant circuits are used extensively in the transmission and reception of radio signals. In fact, the earliest radio receivers were nothing more than a resonant circuit and a small piece of crystal used for detection of radio waves. Resonant circuits are popular in today's receivers because they can be used to "tune-in" various stations.

Think for a moment about the terminology we use when talking about radio. Even people lacking technical knowledge say that stations operate on "frequencies" and that we must "tune" them in. Well, what have we been doing with our resonant circuits? That's right, we've been "tuning" our circuits to different "frequencies" by changing capacitance and inductance. Through a radio's resonant circuit, we can tune around and pick up various stations operating on different frequencies.

RADIO FREQUENCY TRANSFORMERS

We know that we can alter the frequency of a resonant circuit by changing the value of either the inductor or capacitor. However, for practical reasons, in radio receivers (Fig. 6-4) we usually do all our tuning through a variable capacitor. The most popular inductor, on the other hand, comes in the form of a transformer—an *rf* (radio frequency) *transformer.*

We've seen transformers used before to raise and lower ac voltages, but exactly how do they apply in a receiver's resonant circuit? Strange as it may seem, there are actually currents flowing through the air. We call these currents *radio waves* or *electromagnetic waves,* and they act, in a circuit, like alternating current.

We use the transformer, in this application, as a *coupler*—a device that transfers energy without any direct electrical connection. This is very important, because a radio antenna is not at all selective in the current it intercepts. Any energy flying through the air will be picked-up by it and sent into the transformer primary. These currents are then transferred over to the secondary, but since it is part of the resonant circuit, the secondary accepts only the specific frequency the circuit is tuned to.

Radio frequency transformers can be either *loosely coupled* or *closely coupled*. In loosely coupled transformers, the secondary is separated from the primary by a considerable distance. In closely coupled transformers, the primary and secondary are kept in close proximity. Loose coupling results in much better selectivity between received stations than closely coupled transformers. However, since inductive interaction is reduced in this arrangement, the voltage induced in the secondary is much weaker. Figure 6-5 shows a schematic of a typical rf transformer circuit.

PROJECT 18: SERIES-RESONANT DIODE RADIO

This might be the most sensitive radio ever designed (Fig. 6-6), and it may not produce a very large audio output, but look at it this way . . . at least you'll never have to buy batteries or find an ac outlet for it! If we haven't convinced you that radio waves actually contain electrical power, this project ought to prove it. You see, the radio you build from these plans gets its power from only one source—those electromagnetic radiations flowing all around us.

This radio might seem crude, but it works, and if you live in a large metropolitan area you should be able to hear many stations on it. Even in

Fig. 6-4. Radio receivers, such as this one, use a variable capacitor for tuning (courtesy of Panasonic).

77

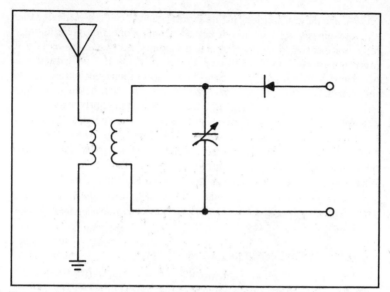

Fig. 6-5. Schematic of a radio circuit employing an rf transformer.

relatively isolated areas, you may be able to pick up a couple stations.
Here's how it works. Signals are intercepted by the antenna, which feeds
the current down a wire to a series combination variable capacitor and
inductor. The end of the inductor should be connected to an earth ground

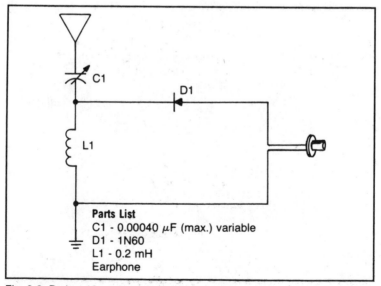

Parts List
C1 - 0.00040 μF (max.) variable
D1 - 1N60
L1 - 0.2 mH
Earphone

Fig. 6-6. Project 18: series-resonant diode radio.

(such as a cold water pipe) for proper operation. The capacitor works, of course, to change the circuit's resonant frequency so different stations can be tuned in. A diode is used for detection and the earphone (any type will do) is for listening.

The antenna for this circuit is very important. We recommend a big one. Actually, the bigger the better. Just use any old wire you can get your hands on and stretch it out as far as you can. If you live in a private house, try stringing it between trees in your backyard. If you're an apartment dweller, either circle the antenna around the ceiling of a room or hang it out the window. Just remember; the longer the antenna, the better the results.

PROJECT 19: PARALLEL-RESONANT DIODE RADIO

The previous project showed a series arrangement used for the reception of radio signals. Here's a circuit using parallel-resonance for the same purpose (Fig. 6-7).

In this circuit, as the previous one, a variable capacitor is used to make the circuit resonant at different frequencies. But that's where the similarity ends. In Project 18, the circuit worked by canceling out reactance at a desired radio station frequency, and letting the maximum current flow at this point. But in a parallel-resonant circuit, as you know, current flows at a minimum when the circuit is in resonance. But if current isn't flowing, how

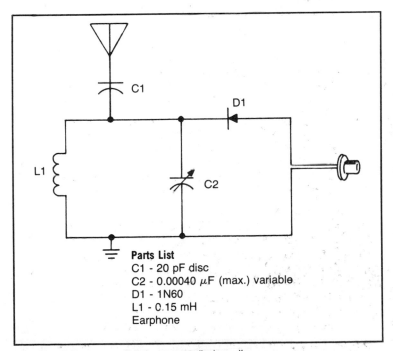

Parts List
C1 - 20 pF disc
C2 - 0.00040 μF (max.) variable
D1 - 1N60
L1 - 0.15 mH
Earphone

Fig. 6-7. Project 19: parallel-resonant diode radio.

can we hear the signal? Easy. Looking at the schematic we can see that when the current comes down the antenna (passing through a small coupling capacitor on the way) it runs smack into the parallel-resonant circuit. Since a signal at the circuit's resonant frequency won't be able to flow through the LC combination, it slides off through the path of least resistance and goes through the diode and earphone and, eventually, to ground. The result is the same as in the earlier project (hearing the radio station), only the method is different.

PROJECT 20: COMBINATION RESONANT CIRCUIT DIODE RADIO

We can make an even better radio by combining both series and parallel-resonant circuits into our receiver (Fig. 6-8). The main advantage here is that by using *two* variable capacitors, we will get better selectivity in rejecting an unwanted station's signal.

In this receiver, we use the series-resonant circuit to tune in the station we want to hear. The parallel-resonant circuit is then used to remove the signal of any interfering station. It works this way. Say we want to listen to a station operating at 880 kilohertz, but we have a very strong station at 660 kilohertz that is leaking through (it can happen, no resonant

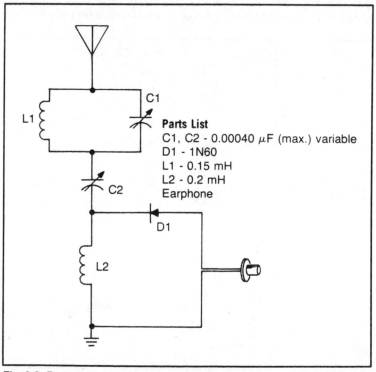

Parts List
C1, C2 - 0.00040 μF (max.) variable
D1 - 1N60
L1 - 0.15 mH
L2 - 0.2 mH
Earphone

Fig. 6-8. Project 20: combination-resonant diode radio.

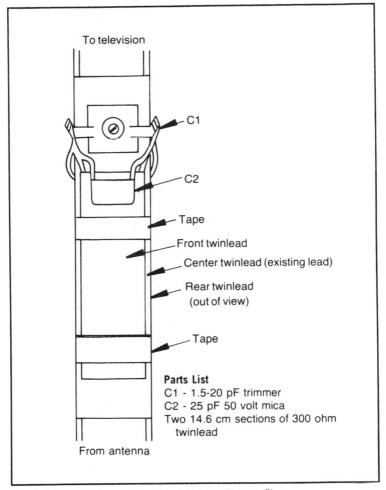

To television

C1

C2

Tape

Front twinlead

Center twinlead (existing lead)

Rear twinlead
(out of view)

Tape

Parts List
C1 - 1.5-20 pF trimmer
C2 - 25 pF 50 volt mica
Two 14.6 cm sections of 300 ohm
   twinlead

From antenna

Fig. 6-9. Project 21: wavetrap television interference filter.

circuit is perfect). In this case, we would tune C2 to a resonant frequency of 880 kHz to let maximum current flow through the diode and earphone at that frequency. Then, to eliminate the interfering station, we would adjust C1 to 660 kHz. Since the parallel-resonant circuit won't let much current flow at that frequency, it gives us an extra stage or rejection, making our reception all that much better.

We call this type of parallel-resonant circuit a *wavetrap*. When we analyze the word, it's easy to see how the circuit gets its name. Radio signals, as we said, are "waves." This circuit works to capture or "trap" radio waves of a specific frequency. As we can now see, resonant circuits can be used to either accept or reject specific frequencies.

PROJECT 21: WAVETRAP TELEVISION INTERFERENCE FILTER

Project 20 showed how a wavetrap can remove radio receiver interference. But did you know we can also use a wavetrap to eliminate television interference? We can, and this project shows how.

Figure 6-9 details a television interference filter you can construct for less than three or four dollars. This little device will help to eradicate interference from CB, mobile radio services, hams, and others. Merely cut two lengths of ordinary 300 ohm television twinlead to a length of 14.6 centimeters. Trim 6.5 centimeters of insulation from each lead from the end of one section. Next, fold the stripped leads toward each other, over the insulation, and solder them together. Now, cut 15 millimeters of insulation from the other end. Repeat this process on the other twinlead section.

After completing work on the two sections, take the mica capacitor specified in the parts list and attach it across the unconnected leads of one of the prepared twinlead segments. Now take the two sections and tape them, in sandwich fashion, around the twinlead running from your antenna to TV. (Note that this circuit makes no direct electrical connection to the TV's twinlead.) Next, bend the lugs on the trimmer capacitor to about a 45 degree angle and attach the leads coming from both twinlead pieces and the fixed capacitor to either lug: lefthand wires to left lug, righthand wires to right lug. The wavetrap is now complete.

Turn on your TV (during a period of interference, of course), and slowly adjust the trimmer until the interference disappears. Since the resonant frequency of this circuit is very precise, if there's more than one interfering station you may need to add more wavetraps along the antenna line.

Chapter 7

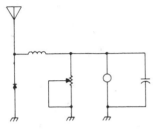

Measuring and Testing Equipment

Information these days really does make the world go around. Electronics plays a large role in this field. Radio, television, computers are all devices created to help deliver and manage information. Yet, in an ironic sort of way, electronics itself is especially vulnerable to a lack of information.

During the course of this book, we've seen many formulas, all designed to tell you what's happening in a circuit under given conditions. Yet, formulas can't always tell the whole story, and even when they can, it's not always convenient to sit down and work out some equations when a decision is needed that very moment. To help us access and monitor various values and quantities, we need to use measuring equipment.

MEASURING VOLTAGE, CURRENT, AND RESISTANCE

Voltage, current, and resistance are the three fundamentals of electronics. We've seen the relationship between these three values expressed earlier in Ohm's law, but we can also measure their action through the use of test equipment.

To measure three values requires three pieces of equipment—true and untrue. To measure voltage, we could use a voltmeter. For current and resistance, an ammeter and an ohmmeter, respectively. These devices do exist, but since most hobbyists, engineers, and technicians need to measure all three values on a regular basis, these instruments have been combined into one—the volt-ohm-milliammeter (vom, for short).

Voms (Fig. 7-1) come in a wide range of prices and styles. All traditional voms, however, have one major feature in common—the large meter that dominates the face. We say "traditional" voms, because a new line of instruments measuring these same values have recently appeared on the

Fig. 7-1. The vom — a most useful meter.

market that don't use conventional meters at all but lighted numerical displays. For now, because they are still in widespread use, let's concentrate on standard voms.

VOM FEATURES

Looking closely at the vom's meter, you'll notice that it is marked with four or five different scales. Each scale represents a different type of measurement—*ac, dc, ohms, dB,* and possibly a *battery check* range used for testing the unit's internal battery (needed for resistance measurements).

Below the meter is a large knob, usually called the *range switch.* The switch selects various functions—ac and dc volts for electrical potential tests, ohms for resistance checks, and dc amps (or milliamperes) for current readings. Beyond that, each setting is broken into various *subranges* for more accurate readings. The number of subranges depends on the sophistication—and the cost—of the vom.

Aside from the main unit, the vom's other major feature is its probe, the device that's physically applied to a circuit to take readings. The probe usually comes in two distinct sections: a pencil-shaped positive lead and an alligator clip used as a ground.

Types of Voms

Voms come in two basic forms: the regular type and the kind employing an FET (field-effect transistor). Why two types? Well, the regular vom has one big disadvantage—a low internal resistance. When measuring circuits with a high amount of resistance, the regular vom has the nasty habit of "loading down" the circuit it's testing. The FET vom sidesteps this problem by containing circuitry that prevents the meter from drawing current from the circuit. The end result, in high resistance circuits, is a much more accurate reading. So if you're going to use a vom to troubleshoot a sensitive or complex circuit, you'll need an FET type. A regular vom, however, works fine for testing individual components and many other common uses.

Using the Vom

There are so many applications for this instrument it would be impossible to describe them all here. Indeed, entire books have been devoted to this single topic alone. Nevertheless, let's look at a few simple vom uses.

Perhaps the vom's most basic task is measuring dc circuit voltage. Unfortunately, it's very easy to incorrectly connect a vom to a circuit. As Fig. 7-2 shows, a vom in the voltmeter mode should always be applied in parallel, across the load. A series connection will cause a reading substantially *below* true circuit voltage. Similarly, it's important to connect a vom measuring current in series (Fig. 7-3). Connect it in parallel and not only will you get a bad reading, but you risk damaging your meter.

Battery testing is another elementary vom use. Most of us, at one time or another, run into battery problems. The most logical way to check voltage would be by touching the appropriate vom leads to either end of the cell; the reading that shows up on the meter should be the voltage being generated. But things don't always work that way. A true voltage test can only be made when a cell is operating under load. A battery can be on its last

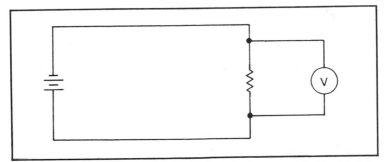

Fig. 7-2. When used as a voltmeter, the vom must be connected in parallel.

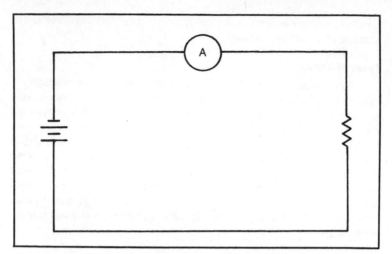

Fig. 7-3. As an ammeter, the vom must be inserted in series—do otherwise, and you may ruin the meter.

legs and still show a reasonable voltage when it's tested out of a circuit. The only way to get an idea of a battery's true condition is to test it while it's actually operating.

To do this, keep the cell in its holder (Fig. 7-4) and place the correct vom leads at either end. Next, switch on the radio (or other battery operated device) and observe the reading on the vom's meter in the dc volts setting. This reading shows the actual voltage being delivered under loaded

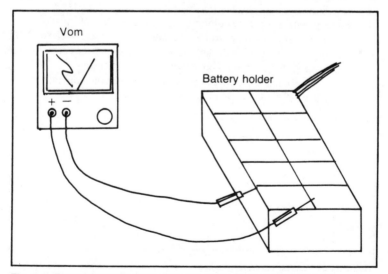

Fig. 7-4. For a reliable reading, batteries must be tested under load.

conditions. If you see one cell that isn't working as well as the others, it should be replaced or removed and charged separately.

Transistor Testing

There are few experiences more frustrating than discovering that the project you just built won't work. Often, the harried builder will pull apart circuit board after circuit board in search of the problem, only to discover that the fault was in a defective transistor.

Another annoying problem is the unmarked, surplus transistor. In these days of rising prices and unwillingness on the part of manufacturers to sell parts in small quantities, hobbyists must necessarily turn to purchasing odd-lot and bargain-basement transistors. While prices for these components are low, the buyer can also be confused by unlabeled and mislabeled parts. Luckily, for determining quality and deciphering unmarked transistors, the vom is a lifesaver.

For instance, an easy way to check for a defective power transistor is to follow the directions in Fig. 7-5. Merely place the probe leads on the terminals indicated, and use your vom (in the *ohms* mode) to check for a high- or low-resistance reading. Don't worry about the actual measurement you get, since the spread between high and low resistance in this test is so outstanding that you'll easily notice the difference. If your readings differ in any way from those of the diagram, there's probably something wrong with the transistor and it should be discarded.

Another transistor check you can make with your vom is to identify and sort bipolar pnp and npn transistors (these devices are described in detail in

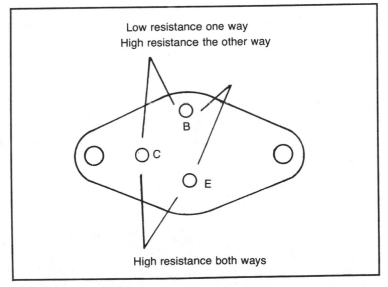

Fig. 7-5. Testing a transistor.

the next chapter). To do this, apply the vom's probe leads to two terminals at a time. One terminal pair will show a high resistance in either direction; these are the emitter and collector terminals. Next, place the negative vom lead on the base terminal and apply the positive lead to either of the remaining unidentified terminals. If you get a low-resistance reading, the transistor is a pnp type, otherwise it's an npn.

The final transistor test is separating the emitter and collector terminals. For this, place the positive and negative probe leads on each of the two unidentified terminals. Follow this by taking resistance readings in both directions. If the transistor is a pnp type, you will obtain a lower resistance reading with the negative probe on the collector and the positive probe on the emitter than you will the other way around. For an npn transistor, the exact opposite is true. A ready-made *transistor checker,* such as the one shown in Fig. 7-6, will also perform all of these tests, and is a very useful piece of test equipment especially if you have many transistors to test.

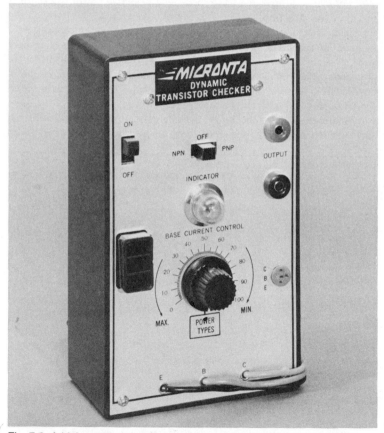

Fig. 7-6. A high-quality transistor checker can test all types of transistors.

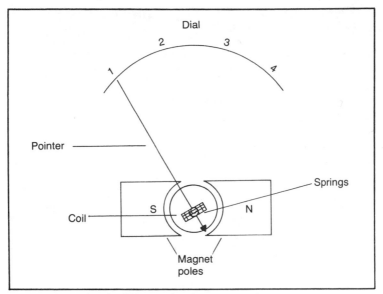

Fig. 7-7. The parts of a D'Arsonval meter movement.

THE D'ARSONVAL MOVEMENT

No, this isn't the name of any political organization seeking increased rights for their people, but it is the name applied to the meter structure found in voms and many other types of equipment. D'Arsonval meters (named after the man who invented them), are the common meter types you see on many run-of-the-mill electronic appliances. You see them used, for instance, as volume-level meters on home tape recorders, as tuning indicators on FM receivers, and as the meter on most CB radios.

This meter type once again returns us to magnetism as a force in electronics. As Fig. 7-7 illustrates, D'Arsonval meters rely solely on magnetism for their function. The idea is really quite simple. A thin wire coil is attached to a pair of springs. At either end of this arrangement is a magnetic pole—one north, the other south. When current flows through the coil, a magnetic field is created causing it to interact with the magnets, forcing the spring to move. This action drives the pointer (which is connected to the coil-spring combination) up the dial's printed scale. When current is removed, or lowered, the pointer falls back toward zero. The result is a very accurate and relatively fast-moving indicator.

DIGITAL METERS

But after decades of cheap, accurate, trouble-free use, the D'Arsonval meter movement's days may be numbered. Increasingly, electronics is turning toward digital displays to give many different kinds of readings. It doesn't take a degree in electrical engineering for someone to notice that

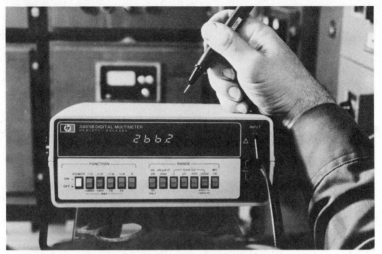

Fig. 7-8. The digital multimeter. Dig those flashy digits!

flashing numbers are replacing dials all over the place—just look at people's wristwatches. Therefore, it's no surprise that the effect is being felt within electronics, too.

The digital display replacement for the vom is known as a *digital multimeter* (dmm) (Fig. 7-8). Actually, the term "multimeter" can also apply to voms (Fig. 7-9), so a dmm might be called a "digital vom." It's all a matter of semantics. A dmm can do everything a vom can with a few extra advantages and some drawbacks. On the whole, dmms cost a bit more than voms, but the price gap has narrowed in recent years. A dmm is easy to read in poorly lit areas (Fig. 7-10) and don't have the parallax problem that afflicts many voms (the apparent difference in reading when viewing a D'Arsonval meter from separate angles). Also, because they don't require a bulky meter movement, dmms can be manufactured to quite compact sizes (Fig. 7-11).

On the other side of this issue, these instruments do have some nagging drawbacks. Dmms usually need an external power source to run those fancy displays, they are more complicated than voms (so may require more repairs), and when encountering pulsating values, their displays can go haywire (a regular vom's pointer would just swing between the two levels).

So which meter is for you? Well, I can't tell you what to buy—that depends upon your specific needs. If you are blessed with money, you might buy one of each type, using a particular meter in the situation for which it is best suited. If you haven't inherited a legacy recently, ask your local electronics dealer to demonstrate each type to you and see which you feel most comfortable with.

Fig. 7-9. An electronic multimeter.

Fig. 7-10. This little digital multimeter is great in dark and cramped locations.

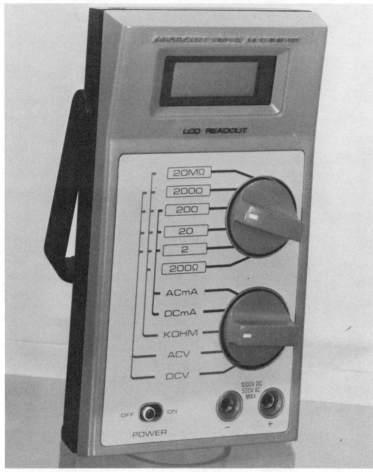

Fig. 7-11. A handheld digital multimeter with LCD readout.

FREQUENCY COUNTERS

Of course, voms aren't the only instruments to feel the impact of digital displays. These little lighted numbers have even managed to inspire a new instrument or two. One such device that's very, very popular is the *frequency counter* (Fig. 7-12).

Frequency counters, not surprisingly, count frequencies. That is, they deduce the ac frequency at a point in a circuit and display it digitally upon their front panel. Now, this isn't to say we didn't have a way of measuring frequencies before counters came along—we did. It's just that these instruments were called *frequency meters*. Frequency meters, for the historical record (nobody uses them any longer), were clumsy, expensive, and hard to work. Counters, on the other hand, are compact, easy to use, and can cost under $100.

Counter Theory and Use

A frequency counter functions by taking its reading at a circuit test point (a place you're interested in getting a frequency measurement), then computing the number of cycles occurring over a precise time interval. It does this by taking the circuit's signal and comparing it with a time signal generated within the counter. This internal signal is very precise, and can determine the test signal's frequency with a high degree of exactness. So let's say that the counter adds up 1,000 cycles during the course of a second. That means there's a signal with a frequency of 1 kHz at that point in the circuit. The end result is a reading on the counter's display for the user to apply.

A counter has many uses. It can tell you the operating frequency of a transmitter and the color burst frequency in a TV; computer clocks, video games, and two-way radios are just a few of the appliances that routinely need counter checks. Anywhere alternating current is present, a frequency counter can be used to determine its frequency (as long as you heed the counter's maximum voltage capability).

Buying A Counter

When shopping for a counter, it's important to note the highest frequency a particular model is capable of reading. To an extent, the counter you purchase will depend on the sort of tasks you want to use it for. If you intend to work on lower frequency radio circuits (like CBs), just about any off-the-shelf counter should do. But if you plan to tackle work on microwave

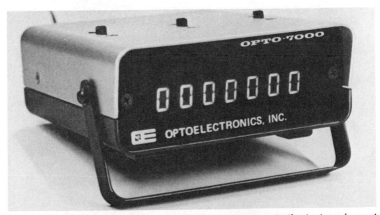

Fig. 7-12. The frequency counter—a relative newcomer to the test equipment field.

equipment, or many types of two-way radios, then a counter capable of measuring up to 600 or even 1,000 MHz may be needed.

As you may expect, the higher a counter can read, the more it will cost. It's very hard to assess a dollar-versus-frequency rule for counter purchases, much depends on the make of the instrument, how many digits you want it to read down to, and so on. You can generally figure about $100 or so for a "regular" counter, another $100 to $200 for one that can measure frequencies up to 500 MHz. Be sure to look around carefully, though. Prices of counters have been quite volatile of late, and you may be able to run into a good deal. Costs can also vary widely between manufacturers, with models from "prestige" companies bearing much higher price tags than those from less well-known manufacturers.

OSCILLOSCOPES

Another important test instrument used by most technicians and many hobbyists is the *oscilloscope* (Fig. 7-13). Of all the test instruments we've mentioned, this unit is by far the most expensive. Even a basic "scope" (as it's commonly called) will cost over $200. If you add some extra goodies to it, the price can easily shoot up to $1,000 or more.

Because of their cost, oscilloscopes are usually owned only by more advanced hobbyists. But it's a shame, because a scope user, with the proper training, can find his troubleshooting work made much easier.

So, for all that money, what can a scope do? Well, it can perform waveform measurements, analyze power supplies, check for stereo amplifier distortion, and a thousand-and-one television repair jobs. Auto repair shops even use this instrument to analyze engine function. In many

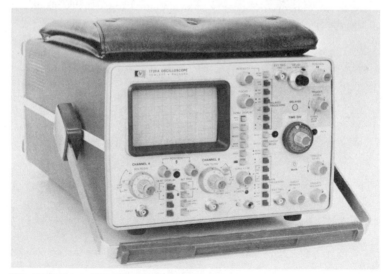

Fig. 7-13. A top-of-the-line oscilloscope.

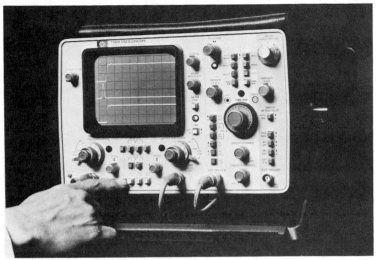

Fig. 7-14. The scope in action.

ways, a scope can do everything voltmeters, ammeters, and frequency counters are capable of, although not as simply. Because of its cathode-ray tube display (similar, in many respects, to a TV picture tube), it may not be the most portable device around, but its many uses can't be disputed. As we said, it's just a shame that they aren't more affordable to the casual hobbyist.

To examine all of the scope's applications is beyond the range of this book. Let's, however, take a quick look at what exactly makes these instruments work. Oscilloscopes, in a nutshell, give us a visual representation of voltage expressed over a precise time interval. The result is a waveform (like the sine wave of ac we saw earlier). Values can be superimposed over each other, providing the user with a way of analyzing very complex signals (Fig. 7-14). By comparing the images on his scope with what he knows should be normal, a hobbyist or technican can determine the exact condition of a circuit. Believe us when we say that oscilloscopes are very important instruments. They're units no well-equipped service bench should be without.

OTHER INSTRUMENTS

What other testing and measuring instruments should be included on a well-stocked bench? Well, after looking at the preceding "universal" instruments, the picture now starts spreading pretty thin. As a rule, we can say that every device and circuit that needs to be tested, has an instrument to do it. For instance, do you need to check the value of an unmarked capacitor? Then the *capacitance meter* in Fig. 7-15 is just the unit for the job. Got a microwave oven you're afraid is leaking? Then use a *microwave leakage detector* (Fig. 7-16). Have a microwave transmitter you want to check the power output from? Then the wattmeter in Fig. 7-17 is for you. On

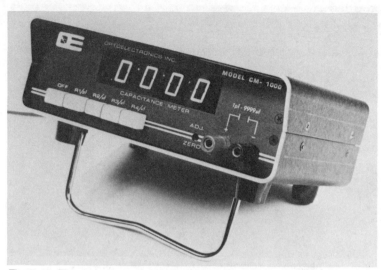

Fig. 7-15. The capacitance meter is a nice addition to any bench.

the less exotic side, you can use the meters in Fig. 7-18 to check the power sent to, and reflected back from, your CB radio's antenna.

PROJECT 22: DIODE-VOM THERMOMETER

In addition to all their other qualities, diodes are also temperature

Fig. 7-16. Cook your dinner, not yourself. This microwave leakage detector checks microwave oven emissions.

96

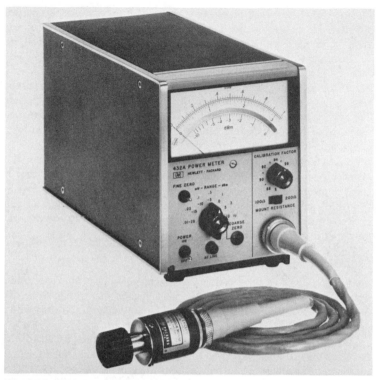

Fig. 7-17. A microwave power meter.

Fig. 7-18. These meters indicate power sent to, and reflected from, a radio antenna.

sensitive when reversed biased (voltage applied opposite to the normal polarity). By using this diode characteristic, along with our ever-handy vom, we can make a simple and accurate vom thermometer. Best of all, we don't even have to worry about buying any specific diode type; since just about all diodes are temperature sensitive, virtually any type will do.

If you set your vom to read ohms, connect a diode between the probe leads (Fig. 7-19), and warm it (or cool it), you'll note the reading change. Try several types of diodes and find the one that provides the greatest range. Make up a chart that tells you how many ohms relate to what temperature and you have a remote-indicating thermometer.

Uses for this project should only be limited by your imagination. For instance, you can place the probe outdoors and monitor the weather in comfort. Just about any task a regular thermometer performs can be done by your electronic thermometer—and with great accuracy. By the way, this project will also work with a digital multimeter. Like they say, the simplest projects can sometimes be the most fun.

PROJECT 23: RF PROBE

Most voms do an excellent job of reading ac voltage. However, radio transmitters operate at frequencies much higher than ordinary ac. To measure these values, your standard vom probe must be replaced with a device known as an *rf probe*.

Rf probes can be purchased at most electronics stores for about $10, but since they're so easy to make, it really doesn't pay to fork out the cash for what a pre-built unit costs. Figure 7-20 shows the few basic parts it takes to make an rf probe, and it shouldn't cost you more than a dollar or two to build. Use your imagination in housing the components. Just about any hollow tubing with a bare wire extending from the end will suffice for the positive lead; an ordinary alligator clip is all you'll need for the ground.

With your rf probe you can adjust radio transmitters (if you have the appropriate FCC license), monitor power output, and adjust antenna matches. If you plan to do any radio work at all, an rf probe is a must.

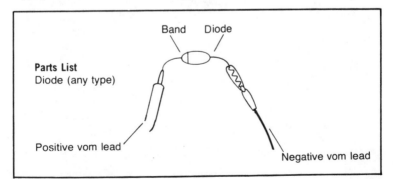

Fig. 7-19. Project 22: diode-vom thermometer.

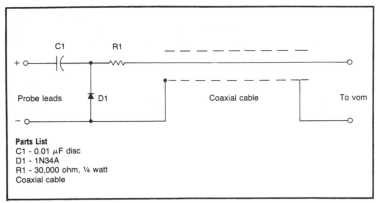

Parts List
C1 - 0.01 µF disc
D1 - 1N34A
R1 - 30,000 ohm, ¼ watt
Coaxial cable

Fig. 7-20. Project 23: rf probe.

PROJECT 24: TRANSMITTER DIGITAL READOUT

Besides giving us very accurate frequency readings for testing purposes, the frequency counter can also be adapted to a more aesthetic use. Whether on an amateur or CB radio, nothing looks snappier than a digital frequency readout. Unfortunately, when purchased as an option, this feature can run into quite a few bucks. If you already own a counter, this circuit (Fig. 7-21) can be added to it for as little as a couple of dollars.

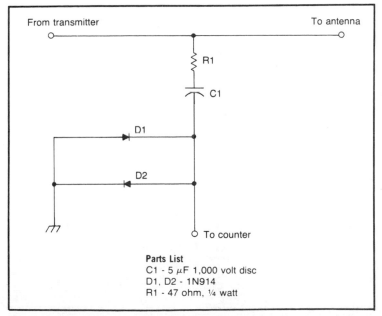

Parts List
C1 - 5 µF 1,000 volt disc
D1, D2 - 1N914
R1 - 47 ohm, ¼ watt

Fig. 7-21. Project 24: transmitter digital readout.

Of course, for this dramatic price reduction, there are a couple of drawbacks. First, this unit will make the counter display the frequency on transmit only. Second, a carrier is needed to make this circuit work, so it is only suitable for AM or CW (morse code). It will not work on single-sideband transmissions. But, still, at this price, we think it's a bargain.

In the past, a common way to achieve this same result was to attach a small whip antenna to the counter. While this worked, its appearance was rather objectionable, the antenna tended to make many counters physically unsteady, and the results were often only marginal. This circuit, on the other hand, is unobtrusive and won't detract at all from either your counter's or room's appearance.

Packaging the circuit is no problem. You can place it inside your transmitter, counter, or in a small cabinet of its own. Be sure to use a shielded cable between the circuit and the counter, if mounting it externally. This device will work up to transmitter power levels of 200 watts without damage.

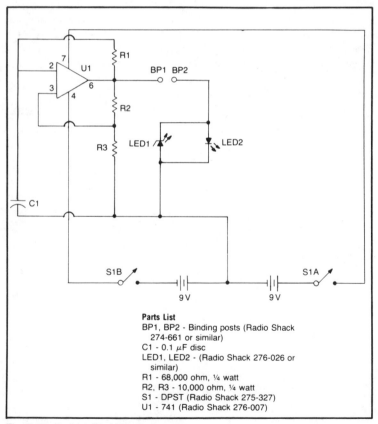

Parts List
BP1, BP2 - Binding posts (Radio Shack 274-661 or similar)
C1 - 0.1 μF disc
LED1, LED2 - (Radio Shack 276-026 or similar)
R1 - 68,000 ohm, ¼ watt
R2, R3 - 10,000 ohm, ¼ watt
S1 - DPST (Radio Shack 275-327)
U1 - 741 (Radio Shack 276-007)

Fig. 7-22. Project 25: LED diode tester.

PROJECT 25: LED DIODE TESTER

Earlier we showed you how to check for defective transistors through the use of a vom. Now, here's a project that will let you analyze the condition of diodes. We think this project does it in a very "flashy" way, too (pardon the pun).

The schematic in Fig. 7-22, shows how you can put together such a tester in just a few hours from commonly available parts. And, if the project is easy to build, its use is even simpler. Just connect the diode in question between the two binding posts. If one LED glows, you've got a functioning diode. Should both LEDs glow, the diode is shorted. Finally, if neither LED lights up, the diode is open. A shorted or open diode, of course, is suitable for installation in a trash can.

Ah, but this tester does even more. Do you have some diodes with obscured or missing bands so you can't tell the anode from the cathode? Well, whichever LED glows with a good diode, indicates which end is the cathode. Use this circuit, and you'll never install a diode backwards again. Unless you're careless, of course. We haven't designed a human carelessness tester yet, but we're working on it.

Chapter 8

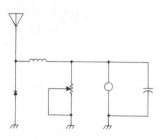

Semiconductor Devices

The vast majority of components and concepts we've studied so far have been around for ages. In some cases, a hundred years or more. But semiconductors are a new branch of electronics, with fresh advancements being made in their technology even as you read these words. It's a dynamic field, full of promise and hope. Semiconductors may someday help to solve our energy crisis, assist manned expeditions to the stars, and help to raise the quality of life right here on earth. And why not? When you consider the awesome advances semiconductors have already given us—from home computers to calculator wristwatches to space shuttles—there's every reason to believe they'll be blazing the way to a better life for us in the future.

SEMICONDUCTOR BACKGROUND

Even if you've never looked into electronics in a serious way before, chances are you've heard about semiconductors. Maybe you're ahead of us and already know what the word stands for, or perhaps you think it is a reference to a sawed-off musical leader. Whatever the case, "semiconductors" seem to be on everyone's mind—especially those "financial wizards" who ride the stock market up and down, cashing in on small electronics companies with new ideas.

Many semiconductors are born in Northern California's "Silicon Valley," that small area near San Francisco where most of the nation's semiconductor industry is located. Quite a few of these multi-billion dollar firms weren't even around a decade or so ago. Others that were, were active in other areas of electronics. Why the big semiconductor boom? To find the answer, we have to look at the change that has taken place in electronics in just the past few years.

You might have noticed by now that we haven't mentioned a component that up until a decade or so ago was very widely used—the *vacuum tube*. And we won't be talking much about the tube, either, except in these few lines. The fact is, the vacuum tube, except for a very few specialized purposes, is dead—gone away with the buggy whip and whalebone corsets. It was the semiconductor that did it in.

At the same time the tube was looking for some shelf space at the Smithsonian Institution, people discovered that the semiconductor, in its various forms, could not only replace the tube, but create new vistas in electronics as well. This opened the door to pocket calculators, miniaturized radios, and the like.

With all these uses for semiconductors coming along, an industry to make these devices sprang up almost overnight. An industry that would not only manufacture semiconductors to replace tubes, but one that would also supply these devices to eager companies who were finding new applications for them all over the place.

WHAT'S A SEMICONDUCTOR?

Why did semiconductors replace tubes in the first place? And why, in any event, did we ever use tubes to begin with? The answer lies in control. We now know of a number of ways to control electron flow: resistance, capacitance, inductance, etc. In the past, we used tubes to control the direction of current flow, and to actually *increase* that flow—an action we haven't seen before. Unfortunately, tubes had their drawbacks—many drawbacks. They were big, needed a lot of power, and eventually wore out. Semiconductors, though, are small, require relatively little power, and last just about forever.

We saw earlier how a diode can control current direction. Well, at one time there were actually tubes called "diodes." Now when we refer to these components, we mean a semiconductor version of the same thing. A diode is the simplest type of semiconductor.

A semiconductor, as you may gather from its name, is a material that neither conducts nor insulates very well. The two most common semiconductive materials are *germanium* and *silicon*. By themselves, neither of these substances are true semiconductors. But, if we *dope* them (add tiny amounts of such substances as boron, aluminum, or gallium), we can make these materials conduct under certain conditions.

By doping with the correct material, we can increase the number of electrons within these substances. Therefore, once impurities have been added, we refer to the altered material as n-germanium, or n-silicon, etc. (as a whole, n-type materials).

Other doping substances can be added to make a positively charged material. Not surprisingly, these are called p-types when completed. The

Fig. 8-1. A semiconductor diode.

atoms in p-type substances have a missing electron, since they have been rendered positively charged. We call these gaps *holes*. Holes can travel through a circuit, carrying current, like electrons—only positively charged.

If we were to hook a section of n-type material to a dc power supply, negatively charged electrons would flow from the supply's negative terminal, through the n-type substance, and into the positive terminal. In a p-type circuit, however, holes would flow from the positive terminal, through the p-type material, and into the power supply's negative pole.

BUILDING A DIODE

A diode (Fig. 8-1), is nothing more than a section of p-type and n-type material stuck together. When placed in a simple circuit (Fig. 8-2), we can see that in a *forward biased* diode (n-type material facing the negative battery terminal) electrons flow from the negative terminal through the n-type material, p-type material, and to the battery's positive terminal. But place the diode the other way, *reverse bias* it, and virtually no current flows. Were we to hook this diode up to a vom, we would notice that when it's reverse biased, resistance would be very high; when forward biased, low. This explains why diodes work to pass current in only one direction.

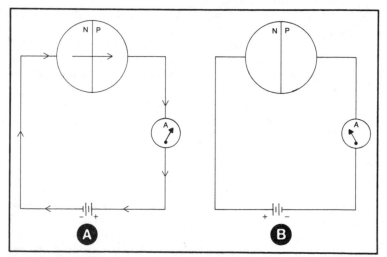

Fig. 8-2. Diode biasing: when forward biased (A), current flow is high; when reverse biased (B), virtually no current flows.

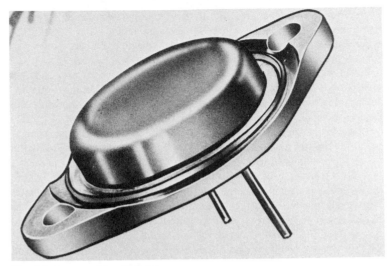

Fig. 8-3. A power-type transistor.

TRANSISTORS

Diodes are great, and we've seen some of their uses, but they don't exhibit the same amplification effect as tubes. To achieve this action, we must use a component known as the *transistor*. Some different examples of transistor forms are shown in Figs. 8-3 and 8-4.

A transistor is made up of three blocks of semiconductor material—either two p-types with an n-type in the center, or two n-types with a p-type

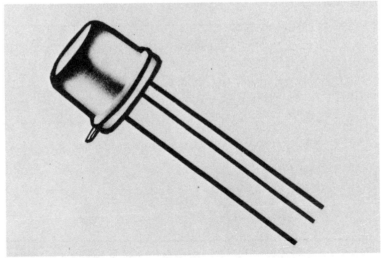

Fig. 8-4. A general-purpose transistor.

center. For clarity's sake, we refer to these configurations as either *pnp* or *npn* transistors. Each of these three semiconductor blocks have a name: the *base, emitter,* and *collector.* (Diode blocks, incidentally, are called the *cathode* (n-type) and *anode* (p-type).)

There is no difference in working action between pnp and npn transistors—only a reversal of circuit voltage polarity. In most applications, transistors are forward biased at the *emitter junction,* while the *collector junction* is reverse biased (junctions being the place of contact between two semiconductor types). As Fig. 8-5 shows, in such an arrangement, the emitter current is large; the collector current small. This is the same result we noted with diode biasing.

But a transistor, of course, uses three elements. Looking at Fig. 8-6, we note that once we connect the emitter, base, and collector, we find only a very small amount of current coming down the base lead. This occurs because most of the current being sent into the emitter is drawn across the base and continues to the collector. Nevertheless, we need the base lead. It acts as a *control,* and if it's disconnected, current out of the collector would drop to nothing.

We also note that a low voltage change in the emitter circuit results in a high voltage change in the collector. This is the *amplifier* action of the transistor at work. In addition, if we increase current at the emitter, current output will also rise at the collector. A pnp transistor works in the same way, but with reversed battery polarities and, of course, holes will be moving, not electrons.

TRANSISTOR AMPLIFIERS

By using transistors, it's possible to make amplifiers for many different applications. Figure 8-7 shows one such arrangement. Here we see the

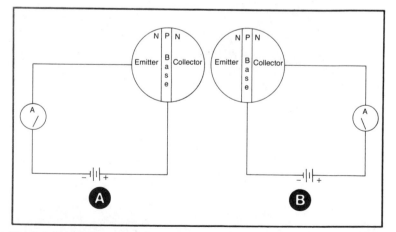

Fig. 8-5. Transistor biasing: forward biased at the emitter junction (A); reverse biased at the collector junction (B).

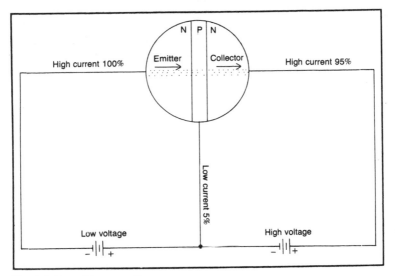

Fig. 8-6. Transistor operation.

transistor set in a *common-emitter* circuit. Our previous example showed a *common-base* arrangement.

In this example, we see that the emitter lead connects between the two batteries. In this configuration, a small voltage change at the base causes a large voltage change at the collector. But the big difference between this circuit and the common-base variety is that a small current change at the base creates a large current change at the collector. Therefore, a transistor connected in this manner is both a voltage *and* current amplifier. Needless to say, this makes it very popular in many electronic uses.

It is also possible to have a *common-collector* set-up (Fig. 8-8). Here, a small voltage change at the base causes an even smaller change at the

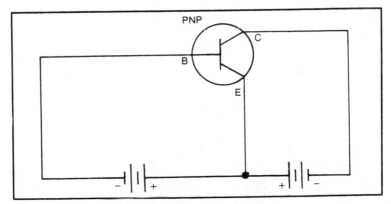

Fig. 8-7. Common-emitter operation.

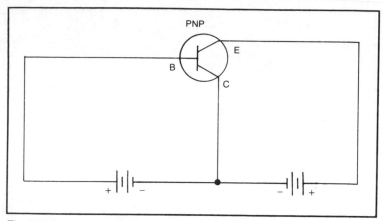

Fig. 8-8. Common-collector operation.

emitter—deamplification. However, a small *current* change at the base makes for a large current change at the emitter. So we can say that this circuit type is only good as a current amplifier.

INTEGRATED CIRCUITS

Like tubes before them, transistors have served as well since their first commercial introduction back in the 1950s. But even transistors are too large for many of today's advanced applications. Transistors, while a big improvement over tubes, still have drawbacks when we must use many of them at once.

We call transistors, capacitors, resistors, and other individual components we've looked at so far, *discrete components.* That is, each part does its own job in its own neat little package. But consider, for a moment, the factor of size. As Fig. 8-9 illustrates, a tube circuit that would have taken up 4 square inches in the 1950s, but would consume only ¾ of a square inch using discrete components. But, by combining all these parts on one *integrated circuit* (IC) chip, the area can be even further reduced to a mind-boggling 2.5 millionths of a square inch.

Yet, the components in an IC chip don't even look like components. We can't pick them up, move them, or even see them without optical aid. For handling, the chip itself must be mounted in a package many times its size (Fig. 8-10), or we'd lose it in a soft wind breeze.

ICs are made through a similar, but much more advanced process, than transistors. A single wafer of p-type or n-type silicon has a doping substance diffused into it. The results are tiny segments of p-type and n-type sections throughout the chip, creating dozens or hundreds of transistors. Each transistor can be accessed and used separately, because each area is like an island unto itself. Besides transistors, other components like diodes, resistors, and capacitors can also be formed on the chip. Resistors are made by

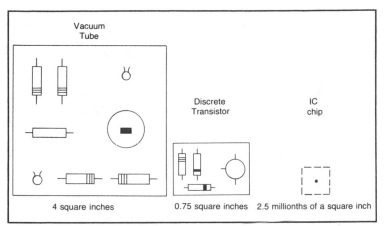

Fig. 8-9. As semiconductors improve, the space they require decreases.

using varying amounts of semiconductor material on the chip base, and capacitors are formed by connecting p-type or n-type substances to a metallized area that acts like a dielectric. It truly takes one's breath away. Just take a look at the microscope view of a full chip in Fig. 8-11, and you'll see what we mean.

LARGE-SCALE INTEGRATION

Would you believe that even conventional integrated circuits are too cumbersome for many tasks? Well, they are, and they're rapidly being supplemented by a technique known as *large-scale integration* (LSI). LSI chips can contain more than 10,000 interconnected electronic elements on a wafer the size of a piece of confetti, as opposed to perhaps hundreds of parts on a small-scale IC. As a matter of fact, these devices are so complicated to

Fig. 8-10. An IC package with the top cut open to expose the silicon chip.

109

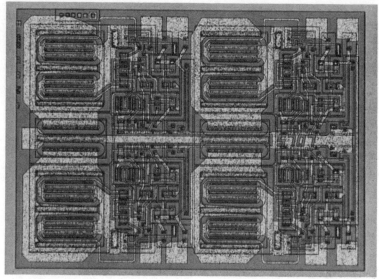

Fig. 8-11. A close-up look at a chip.

design, so intricate, that it takes a computer to do the actual mapping-out and manufacturing work.

An interesting visualization of the complexity involved is shown in Fig. 8-12. In this example, every block in the city of New York represents one element on the pictured chip. To duplicate this job in discrete component fashion, with transistors, would require a half-dozen warehouses to store the circuitry and a medium-sized power plant to run the whole thing. With tubes? Forget it.

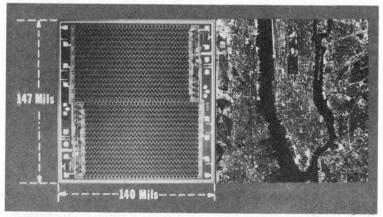

Fig. 8-12. A chip compared to New York City. Every city block is represented by a component on the chip.

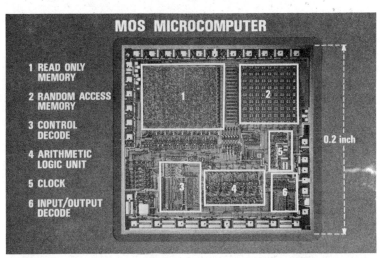

Fig. 8-13. The areas of a microprocessor chip (courtesy of Texas Instruments).

MICROPROCESSORS

LSI technology has given birth to a new form of semiconductor device that is, actually, a computer-on-a-chip—the *microprocessor*. Put another way, a microprocessor is a special purpose or programmable digital computer produced on one or a few LSI circuit chips. A microprocessor, or *microcomputer*, has four basic internal components: a central processing unit (CPU), a *memory* storing and controlling program, memories that hold operation data, and *input/output* circuitry used for linking the system with printers, video displays, sensors, and other outside devices we call *peripherals* (Fig. 8-13).

Microprocessors can be used for two major applications: as a full-fledged computer system for home or business use (Fig. 8-14), or as a system to control a specific appliance or device (Fig. 8-15)—a *dedicated system*. Whatever their applications, and they grow in number every day, all microprocessors hold the advantage of reducing both cost and space.

THE MICROPROCESSOR AS A COMPUTER

Digital computing is based on the speeds at which LSI semiconductors manipulate *bits* of data as electrical pulses. These bits are then organized as binary numbers to represent alphabetic, numeric, and other symbolic information.

A binary number is based on two digits: 0 and 1. The position of these digits in numbers of fixed lengths establish their values. For instance, in a four-digit or *four-bit* system, 0001 represents the decimal number 1; 0010 represents 2; 0011 represents 3; 0100 stands for 4; 0101: 5; 0110: 6; 0111: 7; 1000: 8; 1001: 9, and so on.

Fig. 8-14. Microprocessors make home computers possible.

The benefit derived from using binary numbers is that they can be electronically expressed by two-state electrical values. For example, the digit 1 could be electronically expressed by current on; 0 by current off. If decimal numbers were used, ten different current values would be re-

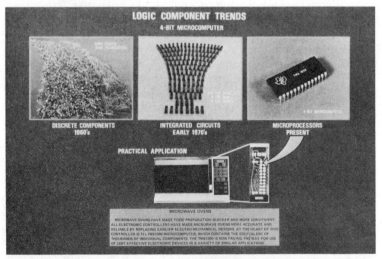

Fig. 8-15. Microprocessor trends in a dedicated system (courtesy of Texas Instruments).

quired. A base ten system may be great if you depend on your fingers, but computers find working with two values much easier.

MICROPROCESSOR IMPACT

Since their introduction in 1970, microprocessors have totally re-shaped the calculator industry (Fig. 8-16). LSI calculators with functions

Fig. 8-16. The pocket calculator—tangible evidence of LSI advances (courtesy of Panasonic).

comparable to electromechanical machines costing around $1,000 in 1969, sell today for less than $10.

Of course, the use of microprocessors in calculators, and such other areas as microcomputing and, say, video discs, are only the most apparent examples. During the 1980s, on-board automobile microcomputers will automatically control engine transmission and brake function, improving performance and reducing pollution. Linked to telephone lines or TV cables, microprocessor-based terminals in our homes will enable us to select news, books, and entertainment for listening, reading, or watching on our TV screens. In addition, we'll be able to program our day's activities so that, for example, lights will be lit and lawns watered at a specific hour. All in all, the microprocessor is quite an impressive device that promises to greatly affect our lives in the years and decades to come.

INSTALLING SEMICONDUCTORS

Up until now, installing a component in a circuit hasn't caused many problems. Ordinary carbon resistors can be mounted in any direction; with capacitors, you might have to note the marked polarity. But, generally, it's been easy enough to figure out how to connect everything together. Semiconductors are different, however. With three leads coming out of transistors, and a dozen or more flowing from the edge of ICs, installation can be a bit tricky.

Diodes are simple to install. A band or dot at the end of the case indicates the cathode. No problem there. How about transistors? Well, transistors come in many different case types. In some instances, you'll see the lead markings molded right into the case. If so, you're all set. Many transistors, however, give no indication as to which lead is which. There are two things you can do in this case: buy a transistor handbook (an item you'll probably want, anyway), or use the transistor checking methods we detailed in Chapter 7. Either way, be sure to get your leads right at installation. It'll save you a lot of trouble later on.

ICs, since they are so complicated, call for even extra care. Pins on these devices are usually numbered beginning at 1, through whatever the last pin is. To find pin 1, look for a small notch or dot in the body of the IC's case (Fig. 8-17). It will either be directly over pin 1, or, looking at the chip head-on, Pin 1 is the first lead on the right-hand side. The remainder of the pins follow in numerical order down the right side, then, back up the left side of the chip so that the last pin is opposite the first. Note: if you see a not-so-deep dot on your chip, in addition to a deeper mark, ignore it. This indentation is a product of the manufacturing process. You're interested in the *deep* dot or notch.

PRINTED CIRCUITS

With the introduction of semiconductor products, also came a new way of connecting components together. Back in the days of tubes, most assem-

Fig. 8-17. An IC showing its notch.

bly was done by using *point-to-point* wiring on a metal chassis. This was an extremely cumbersome method, requiring workers to actually sit in front of the chassis and assemble the unit by hand. The work was slow, tedious, and subject to human error.

With smaller, lighter semiconductors, it became possible to mount parts on boards made out of epoxy or glass laminated with copper. Boards on which components could be mounted and soldered together in a fraction of the time required by older methods. With this great advantage—plus reduced cost—it's no wonder that virtually all of today's electronic equipment is installed on pc boards.

As you can see in Fig. 8-18, all hand wiring is eliminated with pc boards. Instead, metal *traces* connect components together. Although the projects in this book don't require the use of these boards (we recommend beginners use point-to-point wiring on breadboards), you can make your own for future projects by using a kit such as shown in Fig. 8-19.

PROJECT 26: DIGITAL LOGIC PROBE

This project consists of a digital logic probe you can make in your spare time for only a few dollars (Fig. 8-20). In many ways, the logic probe can be considered the digital equivalent to the vom—it's that handy a test instrument. This isn't to call the logic probe a replacement for the vom (each does things the other can't), but they do compliment each other quite nicely. They are also two pieces of equipment no modern electronics hobbyist should be without.

Assembly should be relatively pain-free, and is much like putting together an rf probe for a vom. In this case, you might prefer to enclose the components in a see-through type of tubing (like a large pill bottle), so the LED can be easily read. You may choose to use an opaque tubing, but be sure to cut an opening to view the LED through.

This unit can be used to test all sorts of +5 Vdc circuits. It will tell the user whether voltage is present in a high, low, or pulsating condition. In

Fig. 8-18. A printed circuit board.

other words, it clues you in as to whether a specific digital device is functioning properly in a circuit—a real top-notch troubleshooting aid.

Use of this probe couldn't be simpler. Merely touch the probe lead to the point you want to take a reading. Immediately, the LED will alert you to the test point's condition. If voltage is present in a high state, the LED will flash you a "1." If voltage is low, you'll receive a "0" reading. On the other

Fig. 8-19. A do-it-yourself pc board kit.

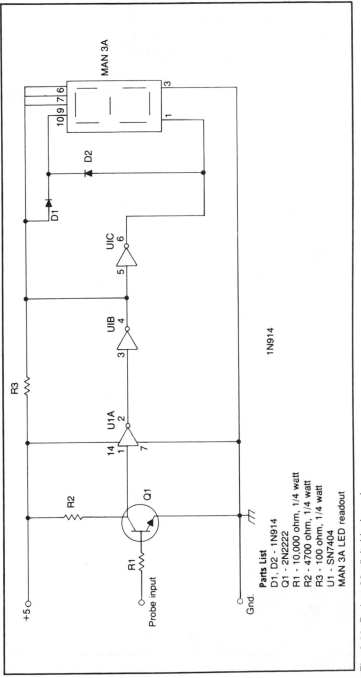

Parts List
D1, D2 - 1N914
Q1 - 2N2222
R1 - 10,000 ohm, 1/4 watt
R2 - 4700 ohm, 1/4 watt
R3 - 100 ohm, 1/4 watt
U1 - SN7404
MAN 3A LED readout

Fig. 8-20. Project 26: digital logic probe.

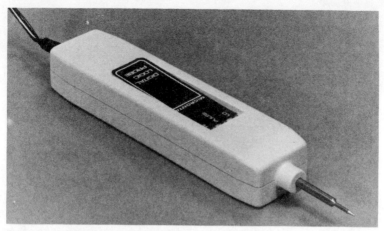

Fig. 8-21. A commercially manufactured logic probe.

hand, if the voltage is pulsating between high and low, you'll see a "P" symbol appear on the LED. With this information at hand, you'll be able to follow exactly what's going on. For instance, if the circuit is supposed to be carrying a high voltage at the test point, and you get a continuous "0" reading, you'll know that things aren't exactly right. Happy troubleshooting. Figure 8-21 shows a commercially available logic probe.

PROJECT 27: IC AUDIO AMPLIFIER

Integrated circuits really are amazing and, in its own way, we think this project helps to prove the point. What we have here is a complete audio

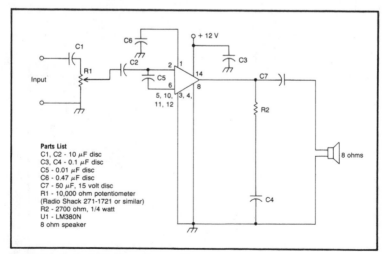

Fig. 8-22. Project 27: IC audio amplifier.

118

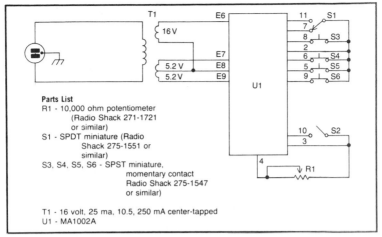

Parts List

R1 - 10,000 ohm potentiometer
(Radio Shack 271-1721
or similar)
S1 - SPDT miniature (Radio
Shack 275-1551 or
similar)
S3, S4, S5, S6 - SPST miniature,
momentary contact
Radio Shack 275-1547
or similar)

T1 - 16 volt, 25 ma, 10.5, 250 mA center-tapped
U1 - MA1002A

Fig. 8-23. Project 28: digital clock.

amplifier that can be built on a board smaller than the size of a cracker (if you place the components close enough together) (Fig. 8-22). By combining many components on a single chip, all that's left for us to do is add a few capacitors and resistors to custom-tailor the circuit for our own use.

This project has dozens of different applications. It will amplify outputs from FM tuners, tape decks, phonographs, intercoms, and so on. If you're into stereo, build a pair of these units, one for each channel. Output is on the order of ½ watt, not high power, but quite sufficient for most uses. All of the parts are easy enough to find, and the IC can be ordered from many retail parts sources.

PROJECT 28: DIGITAL CLOCK

This digital clock project really has everything on one chip (Fig. 8-23). As a matter of fact, with the exception of the switches and transformer, everything is included in one component—even the LED display. It goes without saying that this is a very easy project to assemble. The biggest challenge should be in finding a place to house the board. Our suggestion is a nice wood cabinet, but it's up to you.

The only possible hang-up you may have involves the transformer. As you can see, it's a little on the strange side. Nevertheless, one of the mail order houses is likely to have something at least close to it in stock. Ask when ordering the MA1002A. If not, you might try rewinding an old transformer secondary. It's not as hard as you probably think. Go to your local library and ask for the February 1970 issue of *QST* magazine for a complete guide to this process.

Chapter 9

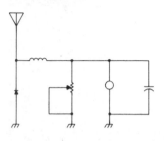

Oscillators

As we've learned by now, ac is a very important facet of electronics. As we've also discovered, ac is generally produced by an ac generator. However, there are applications in electronics where alternating currents are necessary, yet using a generator for these purposes would either be unwieldy at best, or sometimes impossible.

One such area would be in the generation of radio waves. As we mentioned during the chapter on resonant circuits, radio waves are fundamentally ac. Since using a generator to produce these waves would be pretty ridiculous (imagine police officers using handheld radio-generators), a circuit must be available to produce alternating current electronically. We call such circuits *oscillators.*

OSCILLATOR BASICS

All of us are familiar with sound vibrations, such as those created by the plucking of a banjo string or generated by the human larynx. We consider the product of an oscillator as an electrical version of acoustic oscillations.

Oscillators are critical in many areas of electronics. Radios (Fig. 9-1), tape recorders (Fig. 9-2), music organs and synthesizers (Fig. 9-3), and even metal detectors (Fig. 9-4) all make use of this device.

Unlike household ac, which only has a sine-type waveform, we can make oscillators to produce many different waves. Besides sine waves, there are sawtooth, triangular, square, and many other wave categories. If we were to hook an oscillator up to an oscilloscope, we could see the different patterns these waveforms produce (Fig. 9-5).

Fig. 9-1. Radio receivers can use a number of oscillators (courtesy of Panasonic).

OSCILLATOR OPERATION

Just as there are many types of waveforms, we also have a variety of oscillator circuits. But no matter the selection of designs, electronic oscillators all have at least two common elements: an amplifier and a *feedback* method.

We can understand the basic operation of an oscillator by examining the process shown in Fig. 9-6. Here we have a resonant circuit consisting of an inductor and capacitor connected in parallel. When the switch is thrown to connect C1 to the battery, C1 conducts, negatively charging its lower plate. Then, when the switch is moved to connect C1 to L1, C1 discharges and produces a magnetic field around L1. When C1 exhausts its charge, the magnetic field about L1 collapses. But this very act of a collapsing magnetic field induces a voltage of opposite polarity, causing the electrons to reverse their course and charge C1's *upper* plate negatively. With C1 once again charged, this time in reverse, the entire process is repeated. This process keeps repeating over, and over, and over. This is the basic process of oscillation.

But a little thought will tell us that this action can't work in an electronic device without some sort of external support—it would be against the laws of physics. No circuit operates at 100 percent efficiency, so

121

Fig. 9-2. This tape recorder uses oscillators (courtesy of Panasonic).

with each succeeding cycle of oscillation, energy must be lost. After a while the flow of electrons would just gradually fade away, gone through resistance and other factors. A way must be found to replace that lost energy.

TRANSISTORIZED OSCILLATORS

By using a transistor as an amplifier, we can maintain the flow of electrons at a constant level. Here's how it works. Looking at the improved oscillator circuit in Fig. 9-7, we see that current flows through the primary inductor. This forms a magnetic field that affects the secondary inductor,

Fig. 9-3. A music synthesizer employs many waveform outputs for different musical effects.

122

Fig. 9-4. An oscillator is at the heart of this metal detector.

inducing a voltage in it. The induced voltage, in turn, charges C1. When C1 discharges, it does so through the secondary, causing the oscillation process. The loss of energy in this circuit is made up by power from the base and collector circuits.

Another name for our primary inductor is a *tickler coil*. A pretty silly name, but quite apt. What the tickler does is to induce energy to the resonant circuit replacing that lost by resistance. It completes the process known as *inductive feedback*.

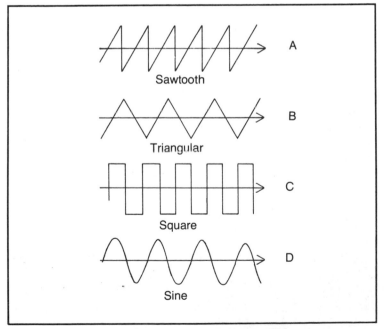

Fig. 9-5. Wave patterns as they would appear on an oscilloscope screen.

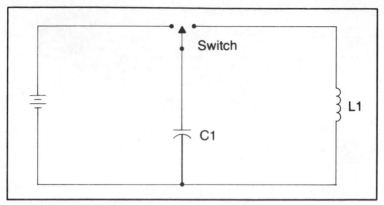

Fig. 9-6. A basic oscillator.

COLPITTS OSCILLATORS

Not all oscillators use tickler coils. There are other ways of feeding back voltage into the resonant circuit. One method is that used by the *Colpitts oscillator* (Fig. 9-8). In this instance, a circuit using only a single coil is tuned to the point of resonance by two series connected capacitors. In the Colpitts circuit, each capacitor shares the ac output from L1. As you can see, the voltage across C2 is fed into the transistor emitter. In this way, the transistor actually supplies its own input, and after C2 builds up enough capacitance, the circuit begins to oscillate. This is an example of *capacitive feedback*.

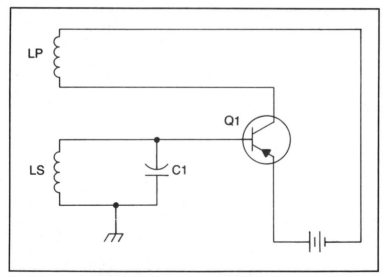

Fig. 9-7. A transistorized oscillator—inductive feedback.

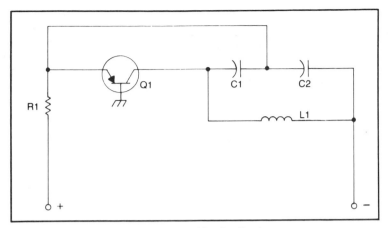

Fig. 9-8. The Colpitts oscillator—capacitive feedback.

HARTLEY OSCILLATORS

Another important oscillator form, used extensively in radio receivers, is the *Hartley oscillator* (Fig. 9-9). Hartley oscillators are easily recognized by the tapped coil in its resonant circuit. In this circuit type, the

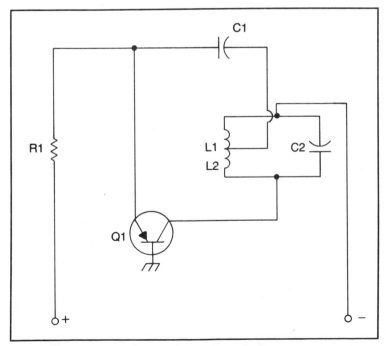

Fig. 9-9. The Hartley oscillator—with its tapped inductor.

125

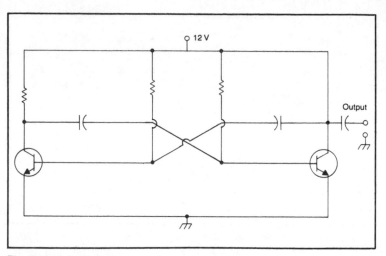

Fig. 9-10. A multivibrator.

inductor is literally split in two sections. The top segment sends voltage to the transistor emitter for amplification; the collector then feeds energy back into the circuit via the bottom coil half.

SQUARE WAVES

So far, all of the oscillators we've been looking at produce only one wave type—the sine wave. However, as we said, there are many different waveforms, so we must have a way of generating these, too.

Perhaps the most common wave pattern, after the sine wave, is the square wave (they're used extensively in television receivers). The most common way of generating this signal is through the use of a device known as a *multivibrator*. A multivibrator is an entirely different sort of oscillator from those we've looked at previously; it falls into the category of a *resistance-capacitance* (RC) *oscillator*.

RC oscillators don't use inductors. As the name suggests, they employ resistors and capacitors (in addition to a transisor or two) to create oscillator function. Looking at Fig. 9-10, we see two identical transistors connected back-to-back. In this arrangement, each transistor is feeding back to the other. Because of this, it operates like a seesaw: as the collector voltage on one transistor rises, the other drops. The end result of this action is a square wave output.

SAWTOOTH WAVES

The third most popular waveform is the sawtooth. Like square waves, this output is commonly used in many TV applications. We can obtain a sawtooth wave by adding something known as a *differential circuit* to our multivibrator. As Fig. 9-11 illustrates, this device uses a resistor-capacitor

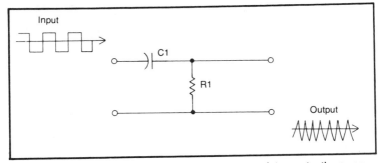

Fig. 9-11. A differential circuit changes square waves into sawtooth waves.

combination. Using these extra components adds a measure of delay to the rise time of the wave; the time needed for the capacitor to gradually charge.

PROJECT 29: AUDIO TEST OSCILLATOR

One way we can put an oscillator to work is by building it into a piece of test equipment. The circuit in Fig. 9-12 is just this type of unit: an *audio test oscillator* (sometimes also called a *function generator.*)

Compared to most of the projects in this book, this one has a fairly complicated circuit—at least in the number of total parts. But things could be even more intricate. If we had to replace the IC with individual components, they would add up to more than 100 extra parts on our project board. As it is, the circuit uses over 20 components. But at least you'll be happy to know that they are all relatively easy to find.

This unit will produce both sine and square waves. Waveform outputs are selected by using jacks J2 (square) and J3 (sine). S1 selects the frequency range, R1 is used to adjust to a specific frequency, and R11 and R12 control square and sine wave levels, respectively. All of these controls should be mounted on your completed unit's front panel for easy accessibility. R1 should be marked on the front panel with a range from 1 through 10. S1 markings should note the various ranges. The circuit's trimmers are installed on the circuit board, and should be touched only during the calibration process.

Calibration itself can be accomplished via two methods. If you own a frequency counter, or have access to one, you're all set. Just feed the tones into your counter and adjust the trimmers until the frequency matches the values indicated on the schematic. Without a counter, you can try calibrating the oscillator by ear to any precise frequency source: piano tuning forks, or another oscillator. But try, if you can, to get use of a counter. After all the work you've put into building the unit, it would be a shame to have it poorly calibrated.

Once you've got the project assembled and tested, it's time to put it through its paces. There are literally hundreds of applications for this oscillator. You can use it to test radio transmitters and receivers,

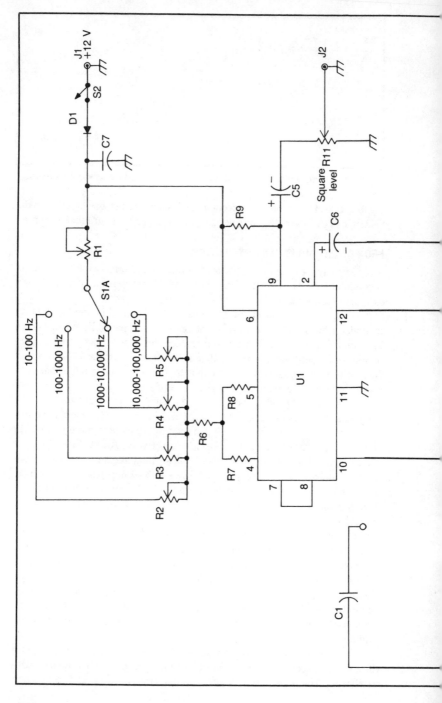

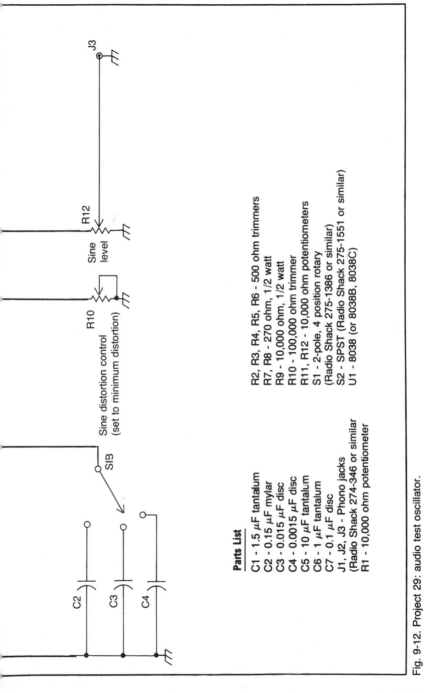

Parts List

C1 - 1.5 μF tantalum
C2 - 0.15 μF mylar
C3 - 0.015 μF disc
C4 - 0.0015 μF disc
C5 - 10 μF tantalum
C6 - 1 μF tantalum
C7 - 0.1 μF disc
J1, J2, J3 - Phono jacks
(Radio Shack 274-346 or similar)
R1 - 10,000 ohm potentiometer

R2, R3, R4, R5, R6 - 500 ohm trimmers
R7, R8 - 270 ohm, 1/2 watt
R9 - 10,000 ohm, 1/2 watt
R10 - 100,000 ohm trimmer
R11, R12 - 10,000 ohm potentiometers
S1 - 2-pole, 4 position rotary
(Radio Shack 275-1386 or similar)
S2 - SPST (Radio Shack 275-1551 or similar)
U1 - 8038 (or 8038B, 8038C)

Fig. 9-12. Project 29: audio test oscillator.

amplifiers, filters, the frequency of other oscillators, and all manner of digital equipment—you can even test the range of human hearing. The deeper you get into electronics, the more uses you'll find for this very handy piece of equipment.

Chapter 10

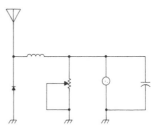

Radio Theory

Radio is, in many ways, the ultimate culmination of all we've been discussing so far. Virtually every bit of theory and every component we've looked at has its application in this field. Because of this, you might think learning about radio is very complicated. Well, it can seem that way. But the beauty of radio, as in all of electronics, is the logical way everything falls into order.

As you should have noticed by now, the various elements of electronics fit together piece by piece—sort of like building blocks. Simple circuits are made more complicated by adding extra parts and stages; each section dovetails into the other, performing its specific task. And in this respect, the field of radio is no different. A radio *transceiver* (combination transmitter/receiver) employs voltage, current, and resistance; uses batteries, resistor coils, a power supply, capacitors, resonant circuits, meters, semiconductors, oscillators, and much more. They all add up to the completed unit. They unify to form the transceiver. Figs. 10-1, 10-2 and 10-3 show some common transceiver types.

PRINCIPLES OF RADIO

As we mentioned back in the chapter on resonant circuits, radio waves are the product of alternating current. The waves themselves, called electromagnetic waves, are actually electric and magnetic fields that travel through space—much like water waves in a pond (Fig. 10-4).

Radio waves have frequency, travel at the speed of light (186,200 miles per second), and are invisible. We know that household ac has a frequency of 60 Hz. Radio frequencies, by comparison, are much higher. Generally, the bottom limit of the *radio frequency spectrum* is considered to be 10 kHz

Fig. 10-1. A typical base station transceiver. This one is designed for amateur use between 50 and 54 MHz.

(10,000 Hz). The uppermost radio signals are transmitted at roughly 300,000 MHz. Figure 10-5 shows a chart of this spectrum.

Of course, no receiver or transmitter could possibly cover all of this territory. Instead, the spectrum is broken into smaller *bands* (Table 10-1). These bands are then divided into even smaller segments, by government authority, and assigned to various *services*. Examples of various radio services are citizens band (CB), amateur (ham), broadcast, public service, business band, etc. Different types of radios are then built for these services, sometimes covering only one or two frequencies, often capable of covering a range of frequencies.

ANTENNAS

We usually think of the receiver as the most important element in radio

Fig. 10-2. A CB mobile transceiver. It operates at about 27 MHz.

Fig. 10-3. Transceivers can also be hand-held, like this CB unit.

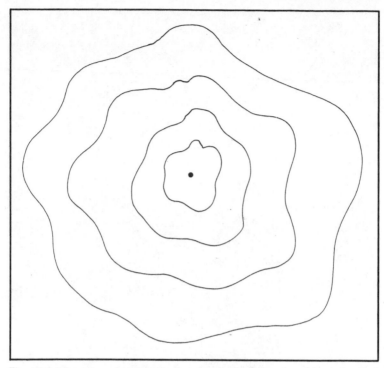

Fig. 10-4. Like waves in a pond, electromagnetic waves travel in circles that become larger and larger.

reception. It's true, receivers are vital. But without an antenna, it would be impossible for us to hear almost any radio signal, no matter how sensitive the receiver. And, without these devices, there would be no way for transmitters to radiate signals, either.

The size and shape of an antenna depends upon a number of factors. For instance, frequency plays a crucial role here. The lower the frequency, the longer the antenna (Fig. 10-6). Furthermore, antennas can be made either *directional* (Fig. 10-7) or *omnidirectional* (Fig. 10-8). Using a directional-type antenna—like a beam (Fig. 10-7)—a transmitting station can concentrate most of its signal to one area. A listener, with a directional antenna, is able to tune out competing stations on the same frequency. Omnidirectional antennas radiate power, and receive signals, to and from all directions with the same level of efficiency. These antenna types are used when a transmitting station wants to cover all areas evenly, or when space prohibits the erection of a directional antenna.

At the transmitter, the radio antenna is decisive in sending radio waves as far as possible. How far, depends to a greater or lesser extent on the band used, the time of day, time of year, how much power the station is sending out, and, of course, on the antenna. Given the right equipment and conditions, there's no limit to the distance a radio signal can cover. Consider, for example, the unmanned space explorers sent to Jupiter and Saturn—all of their pictures were sent back to earth by radio. Even as you read these words, episodes of "I Love Lucy" and "Hazel" are winging their way to the stars for possible consideration by alien civilizations.

On a terrestrial level, radio waves are *propagated* over long distances by skipping or reflecting off of the earth's ionosphere. VLF and LF signals, under ordinary conditions, won't reflect in this manner. Neither will transmissions in the vhf or above category—they pass right through this region and into space. HF signals (sometimes called "shortwave") will skip,

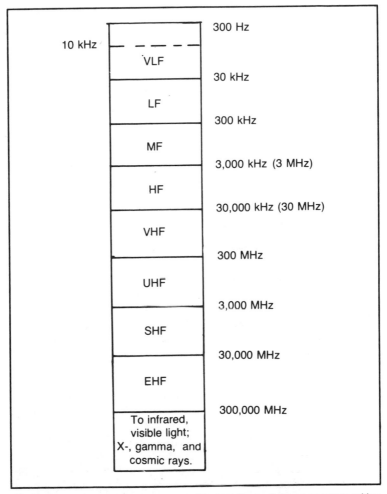

Fig. 10-5. The bands of the rf spectrum. Many radio services may operate with a particular band.

Table 10-1. The Radio Frequency Spectrum.

Band	Abbreviation	Range
Very low frequency	VLF	up to 30 kHz
Low frequency	LF	30-300 kHz
Medium frequency	MF	300-3.000 kHz
High frequency	HF	3.000-30,000 kHz
Very high frequency	VHF	30-300 MHz
Ultra high frequency	UHF	300-3,000 MHz
Super high frequency	SHF	3.000-30,000 MHz
Extremely high frequency	EHF	30,000-300,000 MHz

however, and can provide radio coverage around the globe. Figure 10-9 gives a graphic view of radio propagation.

Satellite communications also work on the theory of reflected signals. Using vhf and higher bands, these satellites intercept signals, amplify them, and return them to a desired point on earth. Because of these sophisticated "relay stations," shortwave is rapidly losing its importance as a long distance communications medium.

THE RADIO RECEIVER

So far, we've been referring often to the radio receiver. Now, let's really take a close look at what exactly goes into one of these units. Fig. 10-10 shows a schematic of a simple receiver we saw earlier. If you built one of these from a project, you know that this circuit is no match for today's state-of-the-art models. But it did, however, show us a lot about the workings of resonant circuits.

Another aspect of radio this device can demonstrate for us is the process of detecting radio waves. When you first saw this circuit type, you may have wondered what the diode was doing in it. We said that it was used for detection, but what does that mean? Aren't diodes only used in places like power supplies to pass current in one direction? True. And that's what it's doing here, too.

Radio signals employ two different waveform types—*audio waves* and *carrier waves*. When transmitting information such as voice or music, audio waves are *modulated* on top of the carrier wave. Unless we do this, conventional radio cannot exist. There is no way to send audio waves through the air alone. They must be "carried" by the carrier.

Back to the diode. As the voltage enters the receiver from the antenna, it's up to the diode to separate the carrier and audio waves and let the beautiful sounds flow. As we remember, the diode is a rectifier—letting current pass in one direction only. When the diode meets the current from the antenna, it slices half of the waveform off, letting only peaks in one direciton (positive or negative) go through (remember our power supply filter?). The resulting half cycles, which average out to follow the audio waveform, leave the diode and enter our earphone.

Of course, at this point, we still have a rather primitive receiving

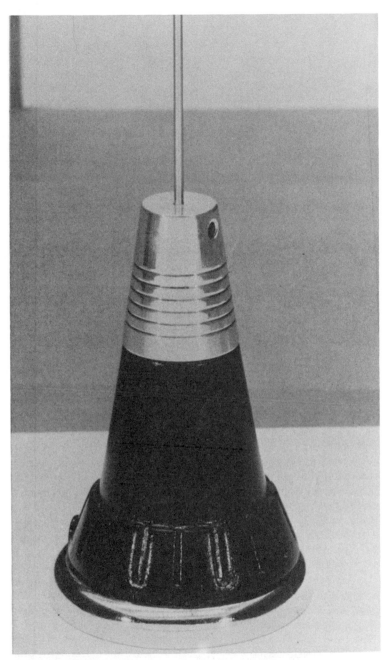

Fig. 10-6. This mobile antenna uses a loading coil at the base to reduce the length of its whip.

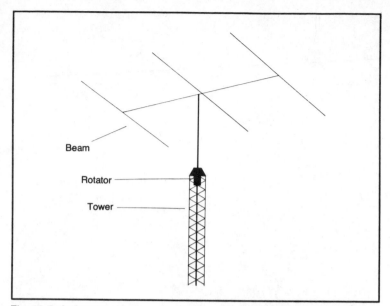

Beam

Rotator

Tower

Fig. 10-7. A directional beam antenna. Sometimes also called a "yagi."

apparatus. A diode is not the world's most sensitive detector, to say the least. To improve the situation, we could use a powered detector stage utilizing one or more transistors and other components. Later on, to raise the output high enough to drive a speaker, we could use one or more audio amplifiers. Only then would we have something approaching the sophistication of a modern receiver (Fig. 10-11).

WHAT CAN WE USE RADIO FOR?

Once we have an understanding of the composition of radio, the next question is what can we do with it? To a large extent, the answer is obvious. We see the applications of radio all around us. Everytime we hop into our car and listen for a helicopter traffic report, turn on our CB to chat with a friend, tune into one of the government's weather stations when danger approaches, or watch our favorite program on television (TV, incidentally, *is* a type of radio—pictures transmitted by radio, to be precise). You may not know this, but TV's full name is *radiotelevision* (Fig. 10-12).

But these are only the most visible uses. Radio can also help us control mechanical devices (Fig. 10-13), and send printed information *Radioteletype*, for example, is used to transmit news stories, business information, and computer data over thousands of miles in the blink of an eye. Newspapers use *radiofacsimile* to receive photographs from the other side of the worlds faster, in many cases, than from across town. Radio also forms the basis of such exotic modes as *radar, radiolocation*, and *microwave data links*.

RADIO EXPERIMENTATION

At this point, you may be interested in learning how you can get in on the action. Unlike the overall field of electronics, radio experimentation is controlled by the federal government through the agency of the Federal

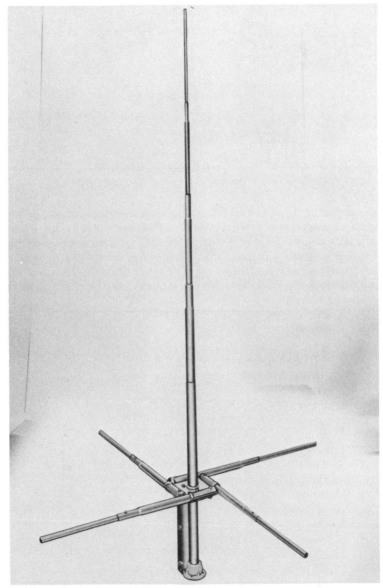

Fig. 10-8. An omnidirectional ground plane antenna.

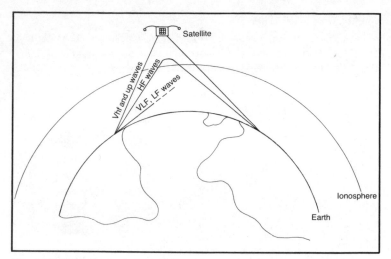

Fig. 10-9. The effects of propagation on various frequency signals.

Communications Commission (FCC). There's a very simple reason for this control. Imagine what would happen if unskilled people were allowed to experiment, at will, on any frequency? They might interfere with military, business, or distress transmissions. They could even do this completely unwittingly, since a misadjusted transmitter can generate signals on frequencies other than those intended. It's for this reason that the FCC tests

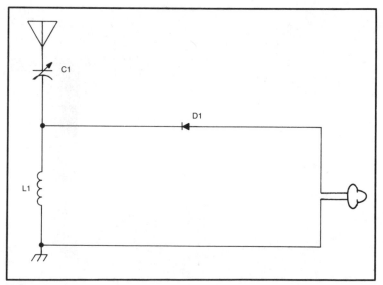

Fig. 10-10. A simple radio receiver.

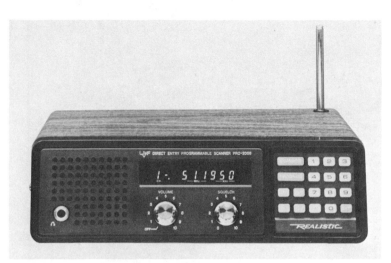

Fig. 10-11. This very up-to-date public service receiver uses a keyboard entry method of selecting frequencies.

individuals for competency before they are permitted to experiment on the air.

Perhaps you've already had a taste of two-way radio communication on citizens band radio. But CB doesn't really fall into the category of radio

Fig. 10-12. A *radio*television (courtesy of Panasonic).

141

Fig. 10-13. Radio control is an important field. Besides controlling toys, it can also free people from dangerous tasks like detonating explosives.

experimentation—it's more like talking on a telephone party line. Only licensed amateur radio operators can build and modify their own equipment, send transmissions over amateur space satellites, and casually chat with friends oceans away.

As we mentioned, you'll have to pass an examination before the FCC will grant you your own operator and station license. But the difficulty of this test has been greatly exaggerated, probably by those who are too lazy to attempt trying it. If you've studied this book diligently, you already know 90 percent of everything you'll need to obtain an entry-level ham license—the Novice class ticket. All you'll have to do is brush up on a few rules and regulations.

In addition to the written exam, you'll also have to pass a basic (five words-per-minute) test on the Morse code. To pass this test, you'll need little more than a recognition ability of the code. People have managed to achieve this status from scratch in less than two weeks. One good way of learning the code is by listening to it in real-life action. Purchase or borrow a good communications receiver like the one in Fig. 10-14, and copy some code for a half-hour or so a night—you'll build up your speed in no time. Even better, you can use your receiver to accompany a transmitter once the feds issue you your license. After you get the hang of receiving code, you'll next want to get a good telegraph key (Fig. 10-15) to practice sending Morse code. Keep up the practice, and you'll be ready to "pound brass" as soon as you get your ticket.

For more information on this fascinating hobby, drop a note to the American Radio Relay League, 225 Main Street, Newington, CT 06111.

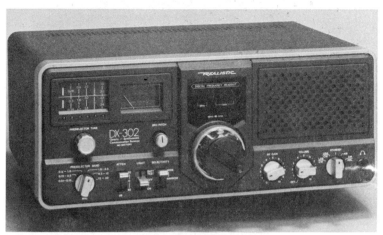

Fig. 10-14. A general coverage receiver with digital readout.

They'll send you a free packet of material along with complete information on how to get started.

PROJECT 30: FIELD STRENGTH METER

If you've ever opearated a radio transmitter, either as a CBer or ham, you know that sometimes it's important to determine whether or not your rig is emitting a signal. Sure, you can always grab the mike and ask someone if he hears you, but that's not very classy. Anyway, what if there is no one listening at the moment?

On the other hand, say you've just installed a new antenna and want to check how well it's radiating. Without a test instrument that can accurately take power measurements from all sides of the antenna, you're in the dark. In either case, a *field strength meter* is what you need.

Fortunately, field strength meters are easy and cheap to construct, and this project shows you how. By following the schematic in Fig. 10-16 you

Fig. 10-15. A telegraph key.

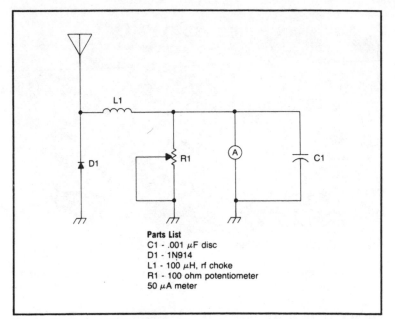

Fig. 10-16. Project 30: field-strength meter.

Parts List
C1 - .001 μF disc
D1 - 1N914
L1 - 100 μH, rf choke
R1 - 100 ohm potentiometer
50 μA meter

should have your meter built and working in no time. By the way, don't pay too much heed to the parts values indicated, anything in the ballpark will do. For an antenna, a random length of wire will work just fine: the longer, the more sensitive the meter. Try mounting the works in an attractive case, and you'll have a test instrument the envy of all.

PROJECT 31: CODE PRACTICE OSCILLATOR

If you'd remember all the way back to Project 9, we showed you how to make a telegraph sounder. Well, that may have been a fine way to learn the Morse code back in the 19th century, but these days something a little more sophisticated is called for.

The schematic in Fig. 10-17, shows a simple code oscillator you can whip together in less than an hour. The parts values aren't at all critical, and can be easily substituted within reason. Just don't swap a capacitor for a resistor, and you'll be okay.

If you're studying for a ham ticket, be sure to have a friend who already knows the code to send some practice text to you. Later, try sending some code yourself to help develop your "fist."

PROJECT 32: IC CODE TRAINER

Here's another project to help you practice the code (Fig. 10-18). We're including it to show you there's more than one way to accomplish the same result.

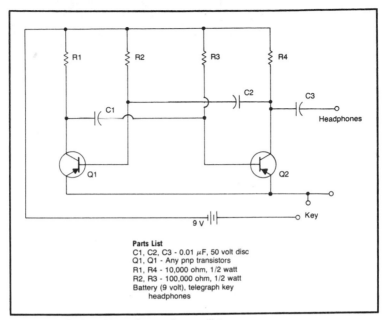

Parts List
C1, C2, C3 - 0.01 μF, 50 volt disc
Q1, Q1 - Any pnp transistors
R1, R4 - 10,000 ohm, 1/2 watt
R2, R3 - 100,000 ohm, 1/2 watt
Battery (9 volt), telegraph key
    headphones

Fig. 10-17. Project 31: code-practice oscillator.

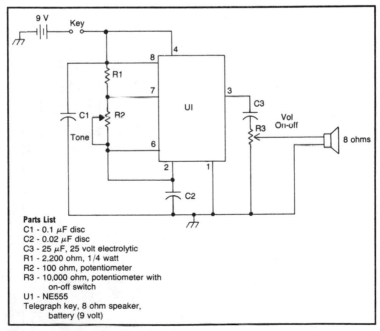

Parts List
C1 - 0.1 μF disc
C2 - 0.02 μF disc
C3 - 25 μF, 25 volt electrolytic
R1 - 2,200 ohm, 1/4 watt
R2 - 100 ohm, potentiometer
R3 - 10,000 ohm, potentiometer with
    on-off switch
U1 - NE555
Telegraph key, 8 ohm speaker,
    battery (9 volt)

Fig. 10-18. Project 32: IC code trainer.

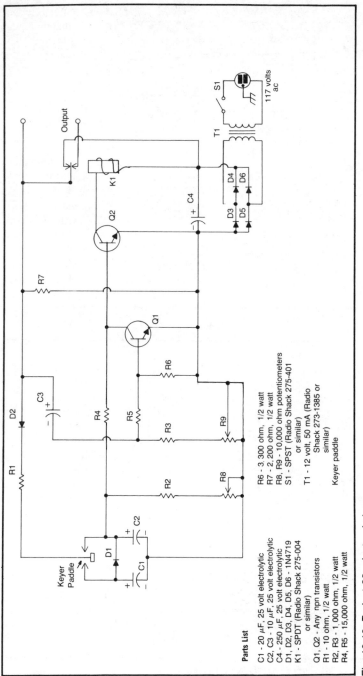

Parts List

C1 - 20 μF, 25 volt electrolytic
C2, C3 - 10 μF, 25 volt electrolytic
C4 - 250 μF, 25 volt electrolytic
D1, D2, D3, D4, D5, D6 - 1N4719
K1 - SPDT (Radio Shack 275-004
    or similar)

Q1, Q2 - Any npn transistors
R1 - 10 ohm, 1/2 watt
R2, R3 - 1,000 ohm, 1/2 watt
R4, R5 - 15,000 ohm, 1/2 watt

R6 - 3,300 ohm, 1/2 watt
R7 - 2,200 ohm, 1/2 watt
R8, R9 - 10,000 ohm potentiometers
S1 - SPST (Radio Shack 275-401
    or similar)

T1 - 12 volt, 50 mA (Radio
    Shack 273-1385 or
    similar)

Keyer paddle

Fig. 10-19. Project 33: electronic keyer.

This project employs the ever-popular NE555 timer chip—probably the most used IC around. It's very versatile. Thanks to this component, and a few extra additions, this unit is a big improvement over the previous code oscillator. There's a volume control, an adjustable pitch, and enough power to drive a small speaker. If you're in a hurry, we suggest Project 31. But if you want a really deluxe code trainer (suitable for classroom use) we advise you take the time and effort and build this one.

PROJECT 33: ELECTRONIC KEYER

Once you've achieved your goal of obtaining a ham license (if you are so inclined), your next step is actual on-the-air operation. Since beginning amateurs (Novice-class licensees) are restricted to using the Morse code, you'll be spending all of your hamming time on that mode. But, as you'll quickly learn, pounding a key can rapidly become very tiring and tedious. To help relieve some of this discomfort, we're presenting you with plans for a device known as an *electronic keyer*.

If you were to buy an electronic keyer in a store, you would probably have to pay at least $30 for a model like the one detailed in Fig. 10-19. By building a keyer yourself, you should save at least $20, possibly even more, depending on how well adept you are at parts scrounging.

The nice part about this keyer is that every single one of its parts is easily available. The actual building is very simple, too. The only part that might cause you some confusion is the "keyer paddle." Actually, all this is is a sort of two-way telegraph key mounted sideways. To send "dits" the key lever is pushed toward the right, for "dahs" you push to the left. Since the dits and dahs come in a continuous string (depending on the lever direction), you'll find the need for repetitive hand movements greatly reduced.

Let's say you want to send the letter Z. Move the key lever to the left for two dahs, then to the right for two dits. Easy. And don't worry about cutting off a dit or dah in the middle. This keyer is "self-completing," meaning that once a code element has begun, it will automatically finish up at the right length, even if you move the lever away prematurely.

You can either buy a pre-built paddle from a store or homebrew your own. In this case, for a change, we suggest buying one. Since a paddle is basically a mechanical item, this is one job you probably can't do as well at home. But, if you insist, you might try buying a stereo phone plug (Radio Shack 274-139 or similar) for conversion to a paddle. Use the ground terminal for a lever, the two positive terminals as your dit and dah contacts. It'll work okay, but it's not very elegant, to say the least.

Chapter 11

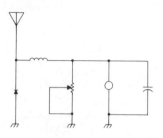

Personal Computers and Space

By now you should have a sound understanding of just what electronics is and why it plays such an important role in our daily lives. Of course, there's more to consider when learning about electronics than just principles and theory of operation. There's a *human* side to the field that's usually overlooked. The final three chapters in this book are an examination of what electronics can mean on a personal level. We'll take a look at what electronics can offer us in the future—both on a hypothetical level and on a practical level—plus a survey of what this field can mean to you as an individual looking for a career. First, let's see how the microcomputers can help man conquer space.

I never knew what welding steel in place was like until I actually tried it. The man at the company's personnel office never promised it would be easy, but he also never said it would be as rought as it was. Imagine, if you can, welding in a zero-gravity situation—nothing to hold onto—everytime you start up the lasertorch, it feels like some space bully is punching you in the chest. It was no picnic, let me tell you.

Another problem was the speed with which they had to recruit workers. That training class instructor went through the stuff so quickly, I barely had time to hear her, much less understand what she said. So what do I think of my personal computer? Well, let me tell you, my friend, if it wasn't for my little digital buddy telling me how and when to do what, me and my lasertorch would probably have been kicked into a permanent orbit around the earth by now.

Transcript Report from a
Spacejourneyman
Pre-fab Welding Sector, X1Z3
July 21, 2014

Does the above fictional excerpt sound a bit farfetched? While space welding operations haven't started on a major scale yet. microcomputers are just about ready to start tackling man's final frontier. Scheduled to ride on an upcoming space shuttle mission is an 11-pound Apple II microcomputer (Fig. 11-1) and, if predictions prove correct, this instrument should only be the first of an entire series of little computers winging their way to the stars.

At first, the idea of using microcomputers in space may sound a trifle silly. After all, can't NASA afford something a bit more substantial? Have budget cuts gone so deep that the government has to buy ordinary personal computers for our astronauts to use? Well, it's a matter of perspective, What we consider "ordinary" computers, just might be the proper tool for the job (Fig. 11-2).

"Tool," a word that we've seen before in this book, is the key word. When our ancestors were first faced with clearing a continent, they too needed tools. For these hardy souls, picks, axes, and hoes were necessities of life. Soon, space pioneers may come to look upon the personal computer as a vital instrument in *their* daily lives.

WHY MICROCOMPUTERS IN SPACE?

Personal computers are ideally suited for space applications because of their small size and considerable computing power. While the micro being carried aboard the space shuttle missions will only be used by

Fig. 11-1. An Apple II microcomputer, much like this model, will ride on-board the Space Shuttle's Space Lab to help scientists conduct experiments. A tentative flight date has been set for early 1984.

Fig. 11-2. Although microcomputers are well-suited for small jobs, it still takes more powerful computers to handle really complex duties. This IBM unit is used to navigate the Space Shuttle.

astronauts and an elite group of scientists, it can be expected that as space travel becomes more commonplace, users of space micros will begin to more closely resemble their earth counterparts.

The micro's greatest promise in space exists in its unique ability to train people on a one-to-one basis. When great masses of people begin migrating to orbiting work stations (perhaps by the beginning of the next century), education will become a formidable problem. For instance, how does one train construction workers, possessing experience in only terrestrial fields, to build massive structures in space? Considering the many specialized tasks that will have to be accomplished, in-class instruction may prove to be outrageously expensive save for only the most general of topics.

By using personal computers, employees will be able to train in the comfort of their own homes, at their own pace, perhaps reporting to a central school only for periodic check-ups and final examinations. Later on, since meeting-space will probably be at a premium, personal computers will allow workers to study new or updated material without returning to earth.

PERSONAL COMPUTER HELPERS

Considering the phenomenal growth in microcomputer technology over the past few years, we can also expect that by the time space industrialization begins in earnest, micros will have developed to an extremely high level of sophistication. With continued advances in both hardware and software (the actual computing "machinery" and its instructions), future space micros will probably use highly developed voice synthesizers and speech recognition systems, allowing users with no computer background to verbally communicate with their micro as if it were a fellow human.

An application of such a microcomputer would be as a "space helper." Since space work promises to be both intricate and hazardous, a computerized assistant could be a real godsend. Imagine, for example, that while working on a deep space repair job, a normally routine procedure suddenly goes haywire. A quick consultation with your personal computer may be able to pinpoint the problem and explain exactly what went wrong.

One important hindrance while working in space is the fact that you can't easily refer to an instruction manual for guidance. By using a personal computer to feed a running commentary into your spacesuit communications system, a complicated repair or service task could be handled much more efficiently. A personal micro might even be able to spot an imminent mistake and give advance warning. Like they say, it's always nice to have a friend when visiting a strange place.

MICRO MONITORING

So far, the microcomputer space applications we've looked at have bordered on the realm of science fiction. But there's one field where micros are not only expected to make a significant contribution, but are already doing so—the area of computer control.

Much space work is made up of necessary but repetitive chores. These jobs include "housekeeping" (assessing and adjusting the environment of space workplaces) and the monitoring of scientific experiments. While housekeeping will be primarily the responsibility of larger computers (Fig. 11-3), individual experiments will, in many cases, be supervised by micros.

The space shuttle's Apple computer is a good illustration of this point. On the shuttle's Space Lab missions, this little computer will be used to monitor the growth of small plants in a zero-gravity situation. Through this experiment, scientists hope to find out why seedlings follow a spiral "helical" growth pattern as they grow. The micro will monitor all phases of the experiment—including growth rates, plant temperature and video imaging—with the collected data being supplied to researchers on earth.

Future microcomputer-controlled experiments, both in the Space Lab and future projects, should only be limited by man's imagination. Entire series of medical, resource exploration, and manufacturing studies can be automatically monitored, leaving humans free to work on more important tasks. Clipboards needn't follow man into space.

MICROCOMPUTERS AS SPACE DESIGN AIDS

The microcomputer's versatility as a design tool for terrestrial en-

Fig. 11-3. Here's a larger computer, but even it's not all that big. This is an isolated view of the Space Shuttle's main computing system. Our good friend the CPU is in the box in the upper-lefthand corner; the memory circuitry is located in the dark-colored box beneath it.

gineers has long been recognized. In space, however, this feature takes on even added importance.

Colossal power-gathering satellites measuring hundreds of miles in diameter, space transportation systems, orbiting colonies—all must be designed and built with the help of computers. While large computers will do the lion's share of the central design work, it's hard to see how any space architectural engineer is going to be able to tackle his assignment without having his own computer close at hand. The sheer mathematics involved in designing such projects would, by itself, exhaust a platoon of scientists. If mankind is indeed going to construct such mammoth projects, portable computers, and lots of them, must be at the fingertips of every single man and woman involved.

WHERE DO WE GO FROM HERE?

This has only been a quick overview of the personal computer's potential in space. Obviously, what you've read here may or may not ever come to pass, but it makes for some interesting conjecture. Only one thing is certain: there will be literally thousands of space applications for micros, just as there are thousands of earthbound uses. All that remains to be seen is exactly what shape these applications will take.

But don't be surprised if, perhaps, somewhere out in the stars, a solitary human will someday form a close relationship with his microcomputer. A flesh and blood and silicon relationship. And together, gently, they will sing: "Daisy, Daisy, give me your answer do. I'm half crazy all for the love of you." Silly? Many people thought the very idea of microcomputers was silly—just 20 years ago.

Chapter 12

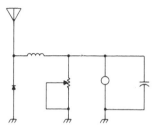

Space Communications and You

For the time being, using a microcomputer in space is nothing more than an interesting future prospect. However, there is a way you can experience the thrill of using space technology without even leaving your home. All it takes is an amateur radio license one grade above the beginner's level and some relatively modest radio gear. Intrigued? Read on.

AMATEUR RADIO'S NEW PHASE

Amateur radio is undergoing a fundamental change that's shaking the hobby down to its very foundations. No, this change isn't springing from the FCC, nor is it coming from any specific alteration in traditional shortwave operation. Nevertheless, from now on, ham radio will never be the same.

What's the cause of this sudden turnabout? Well, it's an upheaval in technology more startling than when hams switched from spark Morse code to voice. It's a revolution that will alter the very frequencies, number of stations, and times of day amateurs will operate. It's the birth of amateur satellite communications.

While hams have been launching their own satellites since the early 1960s, their satellites have been, scientifically speaking, very primitive devices. In the 1960s, for the most part, they were nothing more than aerial beacons hams and shortwave listeners could tune in on their radios. As the 1970s came and went, beacon satellites gave way to two-way birds hams could actually communicate through. But unless you were satisfied with a very brief contact—five or ten minutes, on the average—the satellite would pass out of range before you ever got warmed up. Now, as amateurs enter their third decade of satellite experimentation, the next phase of ham space communications has begun.

HISTORY

OSCAR (Orbiting Satellite Carrying Amateur Radio), has been the name of the ham satellite program ever since the first breadbox-sized unit was launched by Project OSCAR, Inc., in cooperation with the U.S. Air Force on December 12, 1961. Hams have seen eight more OSCARs launched since then. From OSCAR 1, a Sputnik-like craft that beeped the earth with its telegraphic laugh (HI) through the still orbiting OSCARs 8 and 9, which can relay signals over 4,000 miles for up to 25 minutes at a shot, the OSCAR program has united the world's hams while providing tens of thousands of schoolchildren, through classroom demonstrations, with their first taste of space communications.

PHASE III

The course of amateur satellite development can be broken into three distinct eras or "phases." Phase I was the name given to the generation of beacon satellites, Phase II to the latter relay-type units—but it's the new Phase III birds that are sending amateur radio on its revolutionary path. The Phase III OSCARs are as different from their predecessors as a Rolls-Royce Silver Shadow is from a Pinto. Using sophisticated microprocessor technology, these satellites are a totally new species of amateur spacecraft. They are, in effect, scaled-down versions of commercial communications satellites (Fig. 12-1).

One of the changes that makes a Phase III satellite such a radical departure from earlier generations of ham satellites is its very orbit. Instead of circling the earth up to 14 times a day, as earlier OSCARs did, Phase III craft keep a leisurely pace of about two and one-half orbits per 24 hour period. This means that contacts, instead of lasting only a few minutes, can go on for up to nine solid hours—that's longer than many shortwave bands stay open, especially during adverse propagation seasons.

Phase III's usable range will also run rings around its predecessors. Although a highly elliptical orbit will continuously vary its altitude, the satellite will allow every OSCAR equipped station in the Northern Hemisphere to remain within range of each other. But perhaps the best part of working through these new satellites, to the person acquainted with shortwave operation, will be the noticeable absence of negative propagation effects. For the first time in amateur radio history, hams will be able to simultaneously work distant and nearby stations without worrying about the notorious "dead zone" that occurs when shortwave signals "skip" over close-by stations. Also, for the first time, hams will be able to arrange schedules without wondering whether the band will be open at the arranged time. Fretting over propagation charts, sunspot numbers, and solar storms will be worries of the past. An amateur can be sure that if OSCAR is overhead, the contact will be made. That is, if both parties can remember to keep the schedule. Even Phase III OSCAR can't correct defective human memories!

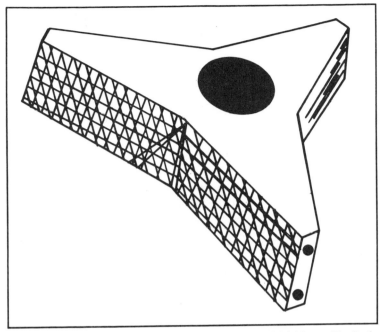

Fig. 12-1. An artist's representation of the Phase III OSCAR satellite that's heralding a new age for amateur radio.

What keeps OSCAR open to so many places at one time is its extreme antenna height combined with vhf-uhf operating frequencies. As we observed in Chapter 10, vhf and higher signals will not reflect off of the ionosphere and return to earth. Therefore, a radio transmission aimed at OSCAR will always arrive at the satellite. OSCAR's antenna height also means that it can actually "see" every station communicating through it. Vhf-uhf's line-of-sight propagation characteristics, so much of a nuisance to earthly TV viewers when a tall building obstructs a signal, actually end up working in OSCAR's favor.

FREQUENCIES

OSCAR uses parts of two separate amateur bands ("passbands," in satellite terminology), to accomplish its mission. The "downlink"— earth receive frequencies—run from 145.850 to 145.990 MHz in the 2-meter band, while the "uplink"—earth transmit frequencies—are 435.150 to 435.290 in the 70 centimeter amateur segment. In other words, a ham transmits on 70 cm and receives signals, relayed through OSCAR, on 2-meters (Fig. 12-2).

Unlike commercial satellites, which limit their customers to specific channels, Phase III OSCAR presents its users a complete 124 kHz

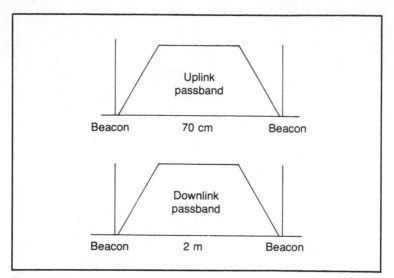

Fig. 12-2. The radio signal goes up on the 70 centimeter band and returns to earth on 2-meters.

passband—with 8 kHz reserved on either end for homing beacons and other special transmission purposes—to spin their frequency dials within. But while operating on OSCAR is generally unrestricted, AMSAT, the non-profit corporation that runs the spacecraft, does ask users and potential users to respect some general operating guidelines. With the vastly superior operating conditions available through the satellite, crowding can be a major problem. Therefore, the satellite's operators have segmented the passband to accommodate the many different amateur transmitting modes.

ACTIVITY

To note the effect these guidelines have on OSCAR's users, let's take a quick tour through the passband and see how the satellite's frequencies are divided. Starting at 145.850 MHz, the lower edge of the 2-meter downlink (Fig. 12-3), the first signal an OSCAR listener will detect is the General Bulletin beacon—the "heartbeat" of the spacecraft. On the beacon, sent alternately in Morse code and radioteletype, is information of critical importance to all OSCAR enthusiasts. Orbital parameters, special user directions, news about satellite operating events—even data on how crowded the passband may be at a given time—are just some of the topics broadcast on this extraterrestrial bulletin board.

Just above the beacon, we next come across three frequencies reserved as "Special Service Channels." The first, Channel L1, is a space set aside for research scientists who will use OSCAR to conduct various experiments approved by the satellite's operators. While most of us may never be able to transmit here, it can make for some pretty interesting listening. Located on Channel L2 is AMICON, the AMSAT International Computer Network. While to the casual listener, the frequency may sound like just so many beeps and buzzes, to the microcomputer-equipped ham the channel will be a meeting place to exchange ideas and, of course, programs. The third special channel is allocated to amateur message handling. When disasters and emergencies strike, Channel L3 is where the action is hottest.

Leaving the SSCs, we next enter the area of prime interest to most hams, the General Communications Band. The GCB is split into three 40 kHz portions, each dedicated to a particular form of amateur communication. Part 1 is the area exclusively reserved for Morse code operation—a place where hams with modest power and equipment can operate without being overwhelmed by stronger stations using voice. Next, comes the center segment—a transitional area where code and voice stations can intermingle—followed by Part 3, which is reserved for voice operations only. Wedged between the code and code/voice segment is a narrow radioteletype band, and straddling the code/voice and voice allocations is a place for slow-scan television transmissions. Overall, any type of operating mode heard on the shortwave bands can also be found on-board Phase III OSCAR.

After the GCB, we again enter a segment of three more Special Service Channels, this time numbered in reverse order. First, is Channel H3, a voice bulletin station frequency providing not only satellite information, but news about all aspects of amateur radio. A unique function of this channel is that amateur radio societies from around the world have selected representatives of their organizations to speak over this frequency and present news of amateur radio in their countries. Channel H2, is devoted to educational purposes, primarily directed at schoolchildren using OSCAR for classroom demonstrations. After H2 becomes the final SSC channel, H1, another international bulletin frequency, this time broadcasting in code and teletype. Tying the ribbon on OSCAR's passband, we reach the Engineering Beacon, a scientific counterpart to the General Bulletin signal at the

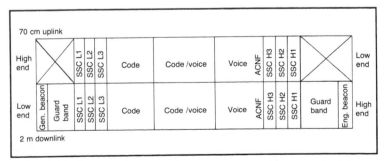

Fig. 12-3. A detailed look at the OSCAR passband. Notice how the links are mirror images of each other, so a signal transmitter on the high end of 70 centimeters comes out on the low end of 2-meters and vice versa.

157

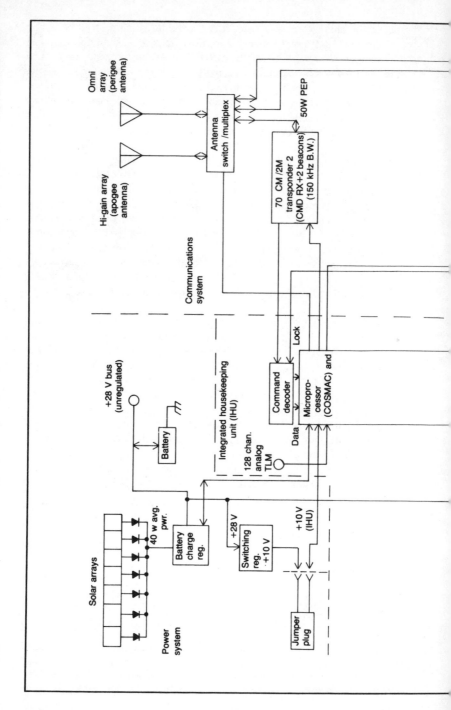

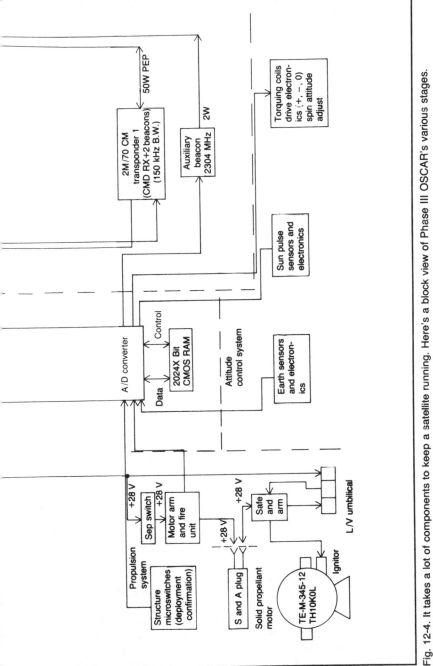

Fig. 12-4. It takes a lot of components to keep a satellite running. Here's a block view of Phase III OSCAR's various stages.

159

start of the passband. Intended mostly for use by ground control stations, this channel generates a steady stream of code and teletype telling about satellite temperature, battery condition, and other housekeeping measurements (Fig. 12-4).

GETTING INTO THE ACT

By this time, if you've developed any electronics interest at all, you're probably itching to know how you can participate in this exciting venture. Well, surprisingly, it's not as difficult or expensive as you may think. In an age where installing a home TV satellite system can cost upwards of $5,000, you can transmit and receive through OSCAR for less than $500—under $200, if you're resourceful.

The first step to getting on-board OSCAR is earning your amateur radio operator's license. Fortunately, thanks to OSCAR's vhf-uhf frequencies, you merely need a Technician-class license—the second lowest amateur class—to possess full satellite operating privileges. The required code speed is a pokey 5 words-per-minute and the theory, while not simple, hasn't prevented many ten-year-olds from getting licenses, either.

If you're already licensed, or on the way, the next step is actually assembling your OSCAR station. Amateur ground stations come in as many forms as shortwave stations, and can be as simple as a shortwave transceiver connected to a vhf-uhf converter, or as elaborate as a custom-designed OSCAR console. But no matter what the size of your wallet or operating space, you can be sure there's a system available to fit your shack or budget. For information on specific types of OSCAR ground stations, you might want to consult *OSCAR: The Ham Radio Satellites,* by Dave Ingram (TAB Books Inc.).

Before leaving the subject of ground stations, there's one more important element to cover—antenna systems. As in every other branch of radio, antennas play a crucial role in determining just how solid a signal you're going to deliver to OSCAR's receiver. And when you consider that OSCAR can fly up to 30,000 kilometers above the earth's surface, the notion of a good antenna system takes on even added importance.

Unfortunately, since OSCAR uses two amateur bands, you're going to need two separate antennas to access the satellite. But, on a happier note, since antenna size decreases as frequency rises, a complete OSCAR antenna system can take up less room and cost less than most CB antennas. For instance, a pair of 12 element OSCAR beams (one for 2-meters, the other for 70 centimeters) will easily mount on any lightweight TV mast or can even be stacked above many existing CB antenna systems.

As with any beam-type antenna, correct aiming is critical for optimum OSCAR performance. While one can "make do" by just pointing an antenna in OSCAR's general direction, ideally—for maximum effectiveness—one would want an antenna system that could follow OSCAR's overhead course. But, for your antenna to track OSCAR on its journey across the heavens, you'll need *two* antenna rotators—a conventional "azimuthal unit, plus an

Fig. 12-5. This azimuthal rotator turns the earth-station antennas around. The elevator rotator tilts the system up and down.

"elevator" rotator that will lift your beams from their horizontal heading and up toward OSCAR (Fig. 12-5).

Actually, the idea of using two rotators shouldn't be much scarier than that of using two antennas. One merely, slips the horizontal boom separating the two beams through a cylindrical opening in the elevator rotator. Then, the elevator's U-bolts are fastened to a conventional mast. From that point onward the system mounts exactly like any other antenna setup. Just imagine your neighbor's faces when they not only see your antenna spinning around, but "nodding" up and down, too!

Obviously, this chapter has only scratched the surface of a field with as much fun, excitment and technical reward as any other electronics specialization. And like any other such endeavor, there's ample room in space communications for people with all levels of skill and expertise. The only important factor in determining your success here, is your desire and willingness to learn. Given such an attitude, in a few years, when amateur space communications becomes the norm instead of the exception, you can relax in your chair, grab your mike and proudly proclaim: "I was a pioneer!"

Chapter 13

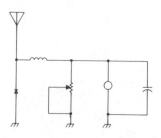

Careers in Electronics

So now we've reached the moment of truth. Just what are you going to do with all of the information you've so recently learned? If you're like most beginners, you'll take this knowledge and begin experimenting in earnest. In time, you'll probably fall into one of the many electronics specializations. You'll become an amateur radio operator, a computer buff, a video enthusiast, or perhaps just an all-around tinkerer. Still, the thought may linger within your mind, "Is it possible to turn all of this into a career? Can I make money from my electronics skills?" The answer, most certainly, is "yes."

CASHING IN ON ELECTRONICS

Perhaps like no other hobby or pastime, electronics experimentation provides its participants with an excellent opportunity to acquire skills immensely useful in the world of time clocks and paychecks. The fact that this training can also be lots of fun is an extra benefit. If you have friends with whom you discuss electronics, the odds are that at least half of them (probably more) are employed in some type of electronics-related field. Now, whether electronics experimentation got these people interested in their jobs or vice versa is subject to debate, but the fact of the matter is that hobby electronics and professional electronics go hand in hand.

What this means to you is that all those hours you spent studying theory and building projects weren't just an intellectual exercise. Depending on how much you've learned, you may be qualified for scores of jobs with only a little extra training. Other, more advanced positions, may require additional formal study; but with your hobby giving you a headstart, you'll probably jump to the head of the class in no time. Whether you're a high school student looking for his first job, a person in mid-life frustrated with

162

his present career, or retired and in need of some extra income, electronics can be the pathway to your financial goal.

TYPES OF JOBS

There are many types of occupations open to job-seeking electronics hobbyists. The Federal Communications Commission, in a published list, has counted more than 65 different career categories covering radio alone. So, contrary to public opinion, not everyone employed in electronics is an electrical engineer (although many are), but run through a wide range of occupations. For purposes of practicality, this chapter will look at five specific fields: broadcasting, television repair, electrical engineering, computing, and two-way radio servicing. While this compilation doesn't cover the entire spectrum of jobs having an electronics connection, it should give you a rough idea of the choices available and show how you can match your skills and interests to a specific career.

BROADCASTING

Although nearly everyone seems to have forgotten by now, electronics experimentation and broadcast radio have very strong links. Stations such as Pittsburgh's KDKA and New York's WQXR (originally, experimental station W2XR), among others, were pioneered by radio amateurs fascinated with the concept of radiotelephone communication. Only when advertisers realized the potential of mass electronic communications, and began pouring into the broadcast radio service, did amateur and professional radio enthusiasts part ways.

Even though the gap between amateur and professional radio is now large, the distance separating the two can be easily bridged. For instance, if you know how to build and operate an amateur transmitter, chances are you can do the same with a commercial rig. And, of course, when it comes to troubleshooting a circuit, it makes no difference whether the rig you're working on is homebrew or commercial. Actually, the greatest barrier separating people in the two services is the FCC. Just as the Commission has a licensing system for amateurs, a similar structure also exists for broadcast technicians. To operate or service any type of commercial radio equipment within the United States, you must hold a commercial license called the "General Radiotelephone License."

Technical Requirements

The examination for this license is rigorous, but not beyond the comprehension of anyone reading this book. While the material included in this test is much too detailed to cover here, your local FCC field office can provide you with the latest syllabus together with an examination time schedule for your area. For more information, you can write to the Federal Communications Commission, 1919 M Street, Washington, D.C. 20554.

While a radiotelephone ticket is necessary to get your foot into broadcasting's door, you'll probably need some outside education, too. This can include technical school, junior college and even, in some lucky cases, on-the-job training. These days, in a switch from the past, many prospective broadcast engineers are also going the four-year college route. An extra benefit here is that most schools offering a broadcasting or communications major will have a radio station and television studio available to students for hands-on experience. Such practice, besides training the student, also gives him the chance to decide which specific area of broadcast engineering to enter.

Compensation

Pay for broadcast engineers is good, from about $9,000 a year for a beginner at a small-town station to over $35,000 for a chief engineer at a "Top 10" market outlet. In a middle-level position, pay ranges between $15,000 and $20,000—somewhat higher at larger, unionized stations and, of course, the networks. In addition, scanning the want ads in *Broadcasting* magazine (available at most larger libraries) will give you a good notion of job opportunities around the country. For more information about careers in broadcast engineering, write to the National Association of Broadcast Employees and Technicians (NABET), 135 West 50th Street, New York, N.Y. 10019.

TELEVISION REPAIR

The receiving end of the broadcasting industry also has ample job opportunities for electronics enthusiasts. Although being a television service technician may not sound as glamorous as working at a radio or TV station, it's still an excellent field in which to make an adequate, steady income (Fig. 13-1).

Fig. 13-1. As technology advances, television servicing opportunities expand. VCR repair and upkeep is a specialty that will take on great importance in the 1980s (courtesy of Panasonic).

Independent Shop or Salaried Employee?

There are two traditional ways to earn money from television repairs. You can either open your own shop and be your own boss, or you can become someone else's salaried employee. If you operate your own shop, there's really no simple way of telling how much income you'll make in a year. You might hit the big time and decide to start a chain of television service centers in shopping malls across the nation, or you could fail and lose everything.

Factors affecting your chances for success will hinge on the size of your community, potential competition and, certainly, your own business sense and technical acumen. On the other hand, working as a technician in someone else's business offers the comforting prospects of a weekly paycheck and job security. As a tradeoff, many TV technicians with an eye toward opening their own business will opt to put in at least five years work as a service shop employee before setting off on their own. This trial period gives the budding entrepreneur time to acquire the skills of the trade while simultaneously learning from his mistakes. The average $11,000 to $15,000 a year such an employee can expect to earn should also help to provide at least some of the nest egg for his eventual business.

Free-lancing

Besides the two ways mentioned above, there's also a third path to making a TV servicing income that's especially appealing to part-timers—free-lancing. With this method you can use your own workbench to service TVs on a small scale in your own home. Of course, you'll have to scrounge your own customers; but by advertising in local newspapers and undercutting the high overhead competition, you should be able to make more than a few bucks. Just be sure to find out *in advance* if your community has any laws regulating TV repair shops.

Shoddy practices by a few TV technicians have forced severe crackdowns by some localities, and you'll want to be sure that you're operating within the law. But whichever route you take, if you're good at troubleshooting and, perhaps, have a year or two of technical school training, TV servicing may be your calling.

ELECTRICAL ENGINEERING

"If you've got the degree, we've got the job." That seems to be the rallying cry for EEs springing from virtually every major newspaper's employment pages these days. Whether you want to be a design engineer, technical writer, or consultant, it's certainly a job hunter's market with companies literally falling all over each other to get qualified applicants (Fig. 13-2). So tight is the current EE job squeeze that many firms in California's "Silicon Valley" have taken to building on-site tennis courts and swimming pools to attract workers.

Needless to say, with some employers even going so far as to fly

Fig. 13-2. Electrical engineering can be a natural offshoot from an interest in hobby electronics.

applicants across the country at no charge just to look over the facilities, many EEs can almost write their own ticket. It's not unusual for someone in such especially desirable fields as microprocessor and energy work having, say, five years' experience, to be earning upwards of $40,000. Not bad. But there is a price to pay for such a rosy picture: and the cost here is education—the more the better.

While a bachelor's degree is a must, the ever-increasing flood of technical information swamping the industry is making engineers holding master's degrees and even doctorates especially sought after. Of course, you'll have to pay to get that education; but with the salaries currently being offered, you should be able to make the money back quickly. Many firms will even refund your education costs as a bonus for signing up. For more information on electrical engineering careers, drop a line to the Institute of Electrical and Electronics Engineers (IEEE), 345 East 47th Street, New York, N.Y. 10017.

COMPUTING

As in electrical engineering, the job situation here is pretty wild. Although employers won't go to quite the same lengths to get most categories of computer people as they do for EEs, the money is tops, the job market very open and, perhaps best of all, you won't have to spend six or more years at college to get started (Fig. 13-3).

Depending on the exact area of computing you're contemplating, you may be able to get away with as little as a year or two at a technical school or junior college. Data processing and computer maintenance are just a couple of the areas open to applicants with as little as a few months of classroom training. Of course, for higher-level positions in such areas as programming or analysis, you'll need a four-year college degree; but, as in electrical engineering, the financial rewards are substantial, so an educational investment can be most worthwhile (Fig. 13-4).

If you're the type of person who enjoys programming his own personal computer, you may be able to enter the computer field without any additional schooling at all—or you can finance your advanced education by working at a part-time computer job.

Depending on your exact skills, many openings exist for computer operators and program troubleshooters with virtually no knowledge of computer electronics but requiring a deep understanding of program languages, something many advanced computer hobbyists seem to possess almost by second nature. Also, most companies specializing in games and other personal computer software are constantly searching for people who can supply them with new and interesting programs (Fig. 13-5). So take a look around! If you regularly operate a home computer, chances are there's a job waiting for you.

TWO-WAY RADIO SERVICING

A long-time career choice for many electronics hobbyists, two-way radio servicing is experiencing something of a renaissance. An explosion in

Fig. 13-3. Data processing is a field open to newcomers with as little as six months of training.

Fig. 13-4. Higher level computer jobs require additional training, but the rewards can be substantial.

two-way radio use, combined with a serious drought of expert technicians, has created an ever-growing demand for skilled people to repair everything from police radios to CB sets.

Since two-way radio technicians—like their broadcasting counterparts—must make transmitter repairs, workers here are also required to hold an FCC General Radiotelephone License. On the brighter side,

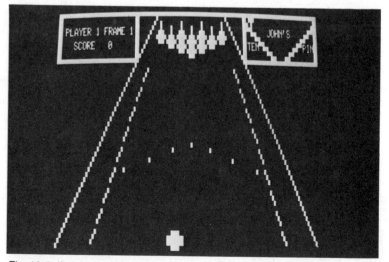

Fig. 13-5. If you know how to program a home computer, you can market your own "software." Here's an example of a video bowling game.

however, your license will prove to most potential employers that you have the skill and knowledge to repair most types of radio gear.

Pay for two-way radio technicians is superb, especially when you consider that a reasonably intelligent electronics enthusiast holding an FCC ticket doesn't need anything more than a basic high school diploma to begin his career. Such a person can expect to make around $10,000 a year to start, with salary prospects going as high as $25,000 for a supervisory position, and even more for repair-department heads for such major employers as the airlines and government.

Speaking of the government, many good opportunities exist within the Civil Service for radio technicians. Federal, state, and local governments are in almost constant need of qualified people to repair and maintain radios for police, fire, safety, and other public services. For the latest line on government jobs in your area, write to your state and local Civil Service departments. If you're interested in working for the federal government, contact the U.S. Office of Personnel Management, 1900 E Street, N.W. Washington, D.C. 20415, for details on Civil Service applications and examinations.

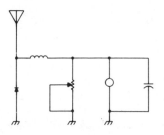

Glossary

ac—Abbreviation for alternating current.

af—Abbreviation for audio frequency.

AM—Abbreviation for amplitude modulation.

ammeter—The instrument used for measuring current flow.

amplify—To increase the magnitude of a signal.

analog—Something that operates with a continuous variation of magnitude.

antenna—The device used to radiate or intercept radio signals.

attenuate—To reduce the effectiveness of a circuit or device.

battery—Two or more cells connected together.

bleeder—A resistor used to drain current from charged capacitors.

bridge rectifier—A full-wave rectifier using four diodes linked together in an interconnecting network.

cable—Usually a bundle of wires used to conduct electricity.

capacitance—Phenomenon that allows a device or circuit to store electrical charge.

capacitor—A component that provides capacitance.

cell—An electrochemical unit that produces dc electricity.

circuit—The path electrons follow.

coil—A conductor that's wound to create inductance.

conductor—A material that allows a free flow of electrons.

current—The flow of electricity past a point.

cycle—The complete path of ac from zero, to positive, to zero, to negative, and back to zero.

dc—Direct current abbreviation.

detector—A device that senses the presence of a signal.

device—A contrivance designed to perform a specific function.

dielectric—An insulating material used between capacitor plates.

dry cell—A cell using a paste-like electrolyte.

E—Voltage symbol.

electromagnet—A temporary magnet made from wire coiled around an iron or steel core.

electromagnetic waves—Waves in space created by an oscillation of electrical charge—radio.

electromotive force—Voltage.

electron—The subatomic particle that carries the negative charge of electricity.

electronics—The branch of physics concerned with the study and control of electrons.

emission—The process of ejecting electrons from a material.

farad—The unit of capacitance.

field—A section of space where a force operates.

frequency—Number of cycles in a given period of time. As in hertz: cycles-per-second.

generator—A device that produces electricity.

germanium—An element used in the production of solid-state devices.

ground—The electrical connection to earth or chassis.

hertz—Cycles per second.

I—Current symbol.

inductance—The characteristic reaction of a coil to magnetic energy.

insulation—A non-conducting material.

K—Abbreviation for kilo.

kilo—Prefix for 1,000, kilovolt—1,000 volts; kilohertz—1,000 hertz; etc.

L—Inductance symbol.

magnet—A device that produces a magnetic field. May be permanent or temporary.

magnetic field—The space surrounding a magnet where magnetic energy exists.

magnetism—The property of an object attracted to magnetic materials.

meter—A measuring instrument.

microphone—A device that transforms acoustical energy into electrical energy.

multimeter—An instrument that measures various electrical quantities—voltages, resistances, etc. Sometimes also called a vom.

negative charge—The charge held by an electron. Also applied to a body with an abundance of electrons.

ohm—The unit of resistance.

Ohm's law—The principle that expresses the relationship between current, voltage, and resistance in a dc circuit.

oscillation—In a circuit, the forward and backward surge of electrons within a conductor.

oscillator—A device that produces alternating current electronically.

oscilloscope—A test instrument used to display ac waveforms on a cathode-ray tube.

P—Power symbol.

polarity—In electricity, plus or minus. In magnetism, north or south.

pole—The terminus of a magnet where magnetic force is concentrated.

potential difference—Voltage.

potentiometer—A variable resistor type.

R—Resistance symbol.

radiation—The emission of energy, such as radio waves.

radio—A wireless communications system using electromagnetic waves.

radio frequency—In general, those frequencies from 10 kHz to 30 GHz used for radio communication.

radio receiver—A device used for the interception and processing of radio signals.

radio telegraph—Telegraphic communications via radio.

radio telephone—Sending voice signals via the radio.

radio wave—Electromagnetic wave.

rectifier—A device or circuit that changes ac into dc.

resistance—The opposition to electron flow.

semiconductor—A material that neither conducts or insulates well.

short-circuit—A low resistance path, often created unintentionally.

signal—Current that carries information.

solid-state—Devices and circuits where electron flow is controlled by specially created solid materials.

spark—A momentary electrical discharge between two terminals.

spectrum—The range of frequencies including radio, visible-light, etc.

static—In radio, atmospheric noise.

subatomic—Particles, like electrons and protons, that are smaller than atoms.

symbol—A letter or design that represents a term or quantity, e.g., E (voltage), ϕ (phase).

transducer—A device that converts one type of energy into another.

transistor—A solid-state device capable of oscillation, amplification and switching. Modern-day replacement for the vacuum tube.

transmission—Moving electromagnetic energy from place to place.

tube—See vacuum tube.

V—Voltage symbol.

vacuum tube—Largely antiquated component used for amplification and other functions by controlling electron flow.

voltmeter—An instrument used to measure voltage.

W—Abbreviation of watt.

wave—A single oscillation of energy traveling through space at the speed of light.

wavelength—The peak-to-peak length of an ac wave.

Appendix A

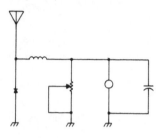

Self-Test

The following self-test has been designed to help you determine just how much information you've retained after reading the technical information in Chapters 1 through 10. If after taking this multiple-choice test you find yourself hazy on certain aspects of the text, be sure to go back and reread the information that may have eluded you on your first try. To help you, the test has been arranged by chapter. There are 20 questions on each chapter.

CHAPTER 1

1. Which of the following *is not* a term relating to electron pressure?
   (a) Voltage   (b) Electromotive force
   (c) Difference of poten-   (d) Power
       tial
2. Which of the following *is* a term relating to electron flow?
   (a) Voltage   (b) Power
   (c) Current   (d) Electromotive force
3. Current is measured in:
   (a) Amperes   (b) Volts
   (c) Watts   (d) Hertz
4. Batteries generate voltage:
   (a) Hydraulically   (b) Chemically
   (c) Through photosynthe-   (d) Through absorption
       sis
5. Carbon cells are known for their:
   (a) Compact size and low   (b) Compact size and high
       cost   cost

(c) Large size and low cost

(d) Large size and high cost

6. Alkaline cells are known for their:
   (a) Compact size and low cost
   (b) Compact size and high cost
   (c) Large size and low cost
   (d) Large size and high cost

7. In a carbon cell, the carbon electrode is:
   (a) Positive
   (b) Negative
   (c) Neutral
   (d) Made of zinc

8. Secondary cells can be charged:
   (a) Only once
   (b) Twice
   (c) Many times
   (d) Cannot be charged

9. Batteries provide:
   (a) An infinite voltage source
   (b) A flow of electrons
   (c) A flow of protons
   (d) Unlimited power

10. AA cells produce:
    (a) 1.5 volts
    (b) 1.5 amperes
    (c) 10.5 volts
    (d) 10.5 amperes

11. The material surrounding cell electrodes is called the:
    (a) Immersion
    (b) Electrolyte
    (c) Compound
    (d) Electroplate

12. What type of current is produced by batteries?
    (a) ac
    (b) dc
    (c) AM
    (d) rf

13. Nickel-cadmium cells are:
    (a) Primary cells
    (b) Secondary cells
    (c) Quadruple cells
    (d) Relatively cheap

14. A lead-acid cell is a:
    (a) Dry cell
    (b) Wet cell
    (c) Moist cell
    (d) Hard cell

15. Lead-acid cells are often found in:
    (a) Pocket radios
    (b) Home appliances
    (c) Hearing aids
    (d) Cars

16. Electrical generators function through use of:
    (a) Chemical action
    (b) Electromotive combustion
    (c) Gravity
    (d) Magnetism

17. The magnets in an electrical generator are:
    (a) Similarly polarized
    (b) Non-polarized
    (c) Oppositely polarized
    (d) Small

18. Which of the following energy sources *cannot* power an electrical generator?
    (a) Oil
    (b) Coal

    (c) Geothermal   (d) Piezoelectricity

19. What type of current is produced by electrical generators?
    (a) ac   (b) dc
    (c) AM   (d) rf

20. Household current operates at a frequency of:
    (a) 110 volts   (b) 110 hertz
    (c) 60 volts   (d) 60 hertz

CHAPTER 2

21. Which of the following *is not* a part of a dc circuit?
    (a) A power source   (b) A load
    (c) A transformer   (d) Copper wire

22. A schematic diagram might be called:
    (a) An electrical road   (b) A nuisance
        map
    (c) A chart of formulas   (d) A list of parts

23. Resistance:
    (a) Speeds electrical flow   (b) Hampers electrical
                               flow
    (c) Has no effect on elec- (d) Is rarely found
        trical flow

24. Resistance is measured in:
    (a) Volts   (b) Coulombs
    (c) Ohms   (d) Watts

25. The components used to introduce resistance into an electrical circuit
are called:
    (a) Resistors   (b) Reactors
    (c) Resisitators   (d) Reluctors

26. Increasing the length of a wire:
    (a) Raises resistance   (b) Lowers resistance
    (c) Has no effect on re- (d) Raises voltage
        sistance

27. The resistance color code for 10% accuracy is:
    (a) Bronze   (b) Gold
    (c) Red   (d) Silver

28. The resistance color code for 5% accuracy is:
    (a) Bronze   (b) Gold
    (c) Red   (d) Silver

29. Which of the following *is not* an Ohm's law formula?

    (a) $I = \dfrac{E}{R}$ (b) $E = IR$

    (c) $R = \dfrac{E}{I}$ (d) $E = \dfrac{R}{I}$

30. In formulas, "I" stands for:
   (a) Inductance (b) Electricity
   (c) Current (d) Voltage
31. In formulas, "E" stands for:
   (a) Inductance (b) Electricity
   (c) Current (d) Voltage
32. Which of the following *is not* a basic foundation of electronics?
   (a) Current (b) Voltage
   (c) Power (d) Resistance
33. Using Ohm's law, say we had a 6 volt circuit with a current of 3 amperes. What would be the resistance value?
   (a) 1 (b) 2
   (c) 3 (d) 6
34. In a circuit with 6 volts and 6 amperes, what is the current value?
   (a) 1 (b) 2
   (c) 3 (d) 4
35. In a circuit with 6 amperes and 12 ohms, what is the voltage value?
   (a) 86 (b) 2
   (c) .5 (d) 72
36. Power is measured in:
   (a) Watts (b) Hertz
   (c) Days (d) Powertrons
37. Which of the nollowing *is not* a formula relating to power?
   (a) $P = EI$ (b) $EI = P$

   (c) $R = \dfrac{E}{I}$ (d) $I = \dfrac{P}{E}$
38. What is the current value of an 8 watt lamp running at 200 volts?
   (a) 0.004 (b) 400
   (c) 0.04 (d) 25
39. Resistors in series:
   (a) Are equal to the sum of the individual resistors
   (b) Have a total resistance that's always less than that of the smallest resistor
   (c) Are equal to one-half their total sum
   (d) Don't work
40. Resistors in parallel:
   (a) Are equal to the sum of the individual resistors
   (b) Have a total resistance that's always less than that of the smallest resistor
   (c) Are equal to one-half their total sum
   (d) Don't work

CHAPTER 3

41. Another name for magnetic force is:

(a) Magnetic structure   (b) Magnetic field

(c) Magnetic resistance   (d) Electromagnetism

42. A naturally occurring magnet is:
    (a) A lodestone   (b) Granite
    (c) Sand   (d) An electromagnet

43. Soft iron is:
    (a) A temporary magnet   (b) A permanent magnet
    (c) A semi-permanent mag-   (d) Non-magnetic
    net

44. Permanent magnets are:
    (a) Hard to magnetize   (b) Easy to magnetize
    (c) Lightweight   (d) Always dark in color

45. Which of the following *is not* a common magnetic material?
    (a) Iron   (b) Steel
    (c) Ferrite   (d) Aluminum

46. Magnets attract:
    (a) Only steel   (b) Only iron
    (c) Only ferrite   (d) Any magnetic material

47. Regarding magnets:
    (a) Opposite poles repel, like ends attract
    (b) Opposite poles attract, like ends attract
    (c) Opposite poles attract, like ends repel
    (d) Opposite poles repel, like ends repel

48. Magnetism helps to produce:
    (a) dc   (b) ac
    (c) AM   (d) rf

49. Electromagnetism is:
    (a) Always important
    (b) Always incidental
    (c) Sometimes important, sometimes incidental
    (d) Never incidental

50. Temporary magnets are:
    (a) Hard to magnetize   (b) Easy to magnetize
    (c) Lightweight   (d) Always dark in color

51. When current flows through a wire:
    (a) It creates a magnetic   (b) It's electrons shrink
    field
    (c) It eliminates the   (d) It lowers the mag-
    field   netic field

52. To obtain the strongest magnetic field possible:
    (a) The wire should be   (b) The wire should be
    coiled   straight
    (c) The wire should be in-   (d) Current should be low-
    sulated   ered

53. Voltage can be induced in a coil when:
    (a) A current moves through it

(b) A magnetic field moves through it

(c) An inductance moves through it

(d) Resistance is encountered

54. Voltage formed in a coil by a neighboring magnetic force is called:
    - (a) Reluctance
    - (b) Inducement
    - (c) Inductance
    - (d) Local voltage

55. Another name for a relay is a:
    - (a) Switch
    - (b) Solenoid
    - (c) Toggle
    - (d) Wired spring

56. In a circuit, inductors behave:
    - (a) In the same way as series or parallel-connected resistors
    - (b) Like a battery
    - (c) In an exactly opposite way as series or parallel-connected resistors
    - (d) Randomly

57. When a coil transfers voltage from one circuit into another we call this action:
    - (a) Mutual reluctance
    - (b) Mutual inductance
    - (c) Mutual inducement
    - (d) Friendly local voltage

58. An electromagnet can be turned on by:
    - (a) Turning on inducement
    - (b) Turning off inductance
    - (c) Turning on current
    - (d) Turning on reluctance

59. Inductors can often be found:
    - (a) In filters
    - (b) In resistors
    - (c) In computer memory chips
    - (d) In batteries

60. The amount of effect between two coils most depends upon:
    - (a) Distance and temperature
    - (b) Angle and resistance
    - (c) Distance and angle
    - (d) Angle and inductance

CHAPTER 4

61. A power supply:
    - (a) Regulates ac
    - (b) Supplies dc
    - (c) Supplies ac
    - (d) Filters dc

62. A power transformer:
    - (a) Raises or lowers ac voltage
    - (b) Raises or lowers dc voltage
    - (c) Lowers ac voltage only
    - (d) Lowers dc voltage only

63. The two transformer coils are called:
    - (a) The first and second levels
    - (b) The top and bottom

(c) The coilers   (d) The primary and sec-
                                       ondary

64. Which of the following *is* the transformer voltage-turns ratio?
   (a) $\dfrac{E_P}{E_S} = \dfrac{T_P}{T_S}$ (b) E_T / T_T

   (c) $\dfrac{E_S}{E_P} = \dfrac{T_S}{T_P}$ (d) T_P / E_S

65. Another name for turns is:
   (a) Circles   (b) Rolls
   (c) Windings   (d) Twirls

66. Which of the following formulas *is* the one for determining power transformer current transfer?
   (a) T_S / T_P (b) T_P / I_S
   (c) $\dfrac{I_S}{I_P} = \dfrac{T_P}{T_S}$ (d) $\dfrac{T_P}{T_S} = \dfrac{I_S}{I_P}$

67. Power transformers always operate at:
   (a) 0% efficiency   (b) 100% efficiency
   (c) Above perfect effi-   (d) Below perfect effi-
       ciency   ciency

68. The rectifier is the part of the power supply that:
   (a) Actually changes ac   (b) Actually changes dc
       to dc   to ac
   (c) Filters ac   (d) Filters dc

69. Which component *can be* used as a rectifier?
   (a) Resistor   (b) Capacitor
   (c) Diode   (d) Inductor

70. A regular full-wave rectifier taps the transformer:
   (a) Halfway through its   (b) Halfway through its
       primary   secondary
   (c) Halfway through its   (d) Halfway through its
       top   first

71. A bridge rectifier has which of the following at all times?
   (a) Full secondary volt-   (b) Full primary voltage
       age
   (c) Full top voltage   (d) Full first level
                                      voltage

72. Pulsating dc is a result of:
   (a) Full-wave rectifica-   (b) Half-wave rectifica-
       tion   tion
   (c) Quarter-wave recti-   (d) Bridge rectification
       fication

73. Unpure dc is said to:
   (a) Sputter   (b) Ripple
   (c) Splatter   (d) Snarl

74. Regulated power supplies:

(a) Supply an unvarying output

(b) Regulate current input

(c) Supply a varying output

(d) Never fail

75. Which component *is not* used as a power supply regulator?
    (a) Zener diode
    (b) Transistor
    (c) Integrated circuit
    (d) Resistor

76. Raising voltage after it leaves the transformer is accomplished by a:
    (a) Voltage multiplier
    (b) Voltage increaser
    (c) Voltage expander
    (d) Voltage increase unit

77. The device mentioned in Question 76 is also:
    (a) A "fast and expensive" way to boost voltage
    (b) A "cheap and dirty" way to boost voltage
    (c) A "slow and clean" way to boost voltage
    (d) A "fast and expensive" way to improve regulation

78. The device mentioned in Question 76 is also:
    (a) A full-wave rectifier
    (b) A half-wave rectifier
    (c) A quarter-wave rectifier
    (d) A bridge rectifier

79. How many rectifiers does a bridge rectifier use?
    (a) 1
    (b) 2
    (c) 3
    (d) 4

80. Which component *can* be used as a power supply filter?
    (a) Resistor
    (b) Capacitor
    (c) Diode
    (d) Limiter

CHAPTER 5

81. Which of the following *is not* a capacitor type?
    (a) Trimmer
    (b) Electrolytic
    (c) Tantalum
    (d) Brundian

82. All capacitors:
    (a) Store resistance
    (b) Store electrical charge
    (c) Store magnetism
    (d) Store protons

83. Which of the following *is not* found inside a capacitor?
    (a) Positive plate
    (b) Dielectric
    (c) Diplatites
    (d) Negative plate

84. The field found inside a working capacitor is called the:
    (a) Electric field
    (b) Magnetic field
    (c) Electromagnetic field
    (d) Power field

85. A capacitor's capacity *does not* depend on:
    (a) The area of its plates
    (b) Its dielectric material
    (c) The distance between its plates
    (d) Its wire lead length

86. Run 250 dc volts through a 100 volt capacitor and you'll get:
    (a) Capacitor breakdown   (b) More capacitance
    (c) Less capacitance   (d) Capacitor let-up
87. Capacitors in series:
    (a) Are equal to the sum of the individual capacitors
    (b) Have a total capacitance that's always less than that of the smallest capacitor
    (c) Are equal to one-half their total sum
    (d) Don't work
88. Capacitors in parallel:
    (a) Are equal to the sum of the individual capacitors
    (b) Have a total capacitance that's always less than that of the smallest capacitor
    (c) Are equal to one-half their total sum
    (d) Don't work
89. The effect of capacitors on an ac circuit is called:
    (a) Capacitive inductance   (b) Resistive reactance
    (c) Capacitive reactance   (d) Inductive reactance
90. The effect mentioned in Question 89 is much like:
    (a) Resistance in a dc   (b) Capacitance in a dc
        circuit   circuit
    (c) Inductance in a dc   (d) Electrons in a dc
        circuit   circuit
91. When inductors are placed in an ac circuit, the effect is called:
    (a) Inductive capacitance   (b) Inductive reactance
    (c) Capacitance reactance   (d) Inductive reluctance
92. Coils are often used to:
    (a) Choke ac   (b) Roughen ac
    (c) Roughen dc   (d) Raise power
93. Which of the following devices *doesn't* use coils and capacitors?
    (a) Television interference filter
    (b) Power line filter
    (c) Air conditioner filter
    (d) Radio transmitter antenna tuner
94. The coulomb is the:
    (a) Unit of quality   (b) Unit of quantity
    (c) Unit of capacitance   (d) Unit of inductance
95. The farad is the:
    (a) Unit of quality   (b) Unit of quantity
    (c) Unit of capacitance   (d) Unit of inductance
96. The henry is the:
    (a) Unit of quality   (b) Unit of quantity
    (c) Unit of capacitance   (d) Unit of inductance
97. Disconnected capacitors:
    (a) Are harmless   (b) Do not store energy
    (c) Are worthless   (d) Can kill

98. You can find 6.28×10^{18} electrons in:
    - (a) A farad
    - (b) A coulomb
    - (c) A henry
    - (d) A volt
99. Capacitor value and ac frequency are important when considering:
    - (a) Reactance
    - (b) Inductance
    - (c) Reluctance
    - (d) Resistance
100. A shorting stick is a type of:
    - (a) Safety device
    - (b) Inductor
    - (c) Capacitor
    - (d) Relay

CHAPTER 6

101. Resonance is:
    - (a) Producing a large stimulus from a small signal
    - (b) Producing a large signal from a small stimulus
    - (c) Producing a large signal from a large stimulus
    - (d) Producing a small signal from a small stimulus
102. In a series-resonant circuit:
    - (a) Current rises to a maximum at resonance
    - (b) Current falls to a minimum at resonance
    - (c) Current remains unaffected at resonance
    - (d) Current splatters at resonance
103. In a parallel-resonant circuit:
    - (a) Current rises to a maximum at resonance
    - (b) Current falls to a minimum at resonance
    - (c) Current remains unaffected at resonance
    - (d) Current splatters at resonance
104. In a resonant circuit:
    - (a) Capacitive and inductive reactances are equal
    - (b) Capacitive and inductive reactances are unequal
    - (c) Capacitive and inductive resonances are equal
    - (d) Capacitive and inductive parts values are equal
105. Resonance is:
    - (a) Greatly affected by the frequency of a circuit
    - (b) Greatly affected by the inductive reactance of a circuit
    - (c) Greatly affected by the capacitive reactance of a circuit
    - (d) a, b, and c
106. Which of the following elements *isn't necessary* to a resonant circuit?
    - (a) Inductor
    - (b) Capacitor
    - (c) Resistor
    - (d) ac
107. A series circuit is:
    - (a) A combination of elements in direct succession
    - (b) A shunt connection of elements
    - (c) Not often used
    - (d) Very simple
108. A parallel circuit is:
    - (a) A combination of elements in direct succession

(b) Components connected across each other

(c) A shunt series connection

(d) Very simple

109. Resonant circuits are used extensively in:
  (a) Radio transmitters
  (b) Radio receivers
  (c) Televisions
  (d) a, b, and c

110. Most radios:
  (a) Use a capacitor for tuning
  (b) Use an inductor for tuning
  (c) Use a resistor for tuning
  (d) Use frequency for tuning

111. The most popular inductor used for radio reception is:
  (a) A power transformer
  (b) An rf transformer
  (c) A simple coil transformer network
  (d) A triple-stage audio transformer

112. An inductor, like the one mentioned in Question 111, is used as:
  (a) A bridge
  (b) An interface driver
  (c) A coupler
  (d) A driver

113. Radio waves, in a circuit, act like:
  (a) ac
  (b) dc
  (c) AM
  (d) FM

114. Another name for radio waves is:
  (a) Magnetoelectric waves
  (b) Electromagnetic waves
  (c) Kilowatt waves
  (d) Radio energy patterns

115. Early radios were nothing more than:
  (a) Tinfoil
  (b) A resonant circuit and a piece of crystal
  (c) A resonant circuit and a couple of tubes
  (d) A dynamotor and an antenna

116. Radio stations operate on:
  (a) Waveforms
  (b) Channels
  (c) Frequencies
  (d) dc waves

117. A radio antenna is:
  (a) Very selective
  (b) Not very selective
  (c) Very powerful
  (d) Always made of wire

118. Which inductor end is connected to the antenna?
  (a) The primary
  (b) The secondary
  (c) The top
  (d) The second level

119. High selectivity is a result of:
  (a) Loose coupling
  (b) Close coupling
  (c) Close driving
  (d) Loose interfacing

120. Inductive interaction in a transformer:
  (a) Stops current
  (b) Lowers current
  (c) Is unimportant
  (d) Is very helpful

CHAPTER 7

121. A vom *does not* measure:
    (a) ac voltage   (b) Resistance
    (c) dc voltage   (d) Frequency
122. A vom's two probe sections are:
    (a) A pencil-shaped positive lead and an alligator clip ground
    (b) A pencil-shaped ground and an alligator clip positive lead
    (c) Two alligator clips
    (d) Two pencil-shaped leads
123. A vom's major disadvantage is:
    (a) Low resistance   (b) High resistance
    (c) High cost   (d) Low cost
124. Vom readings must be taken:
    (a) Under load   (b) Without load
    (c) Often   (d) Only under low power
125. The D'Arsonval meter:
    (a) Is a digital display   (b) Uses magnetism
    (c) Is inaccurate   (d) Is slow moving
126. The digital display replacement for the vom is the:
    (a) FET vom   (b) vtvm
    (c) Digital counter   (d) Digital multimeter
127. Digital meters:
    (a) Are hard to read   (b) Don't have a paral-
                                        lax problem
    (c) Are very, very cheap   (d) Don't work well
128. Frequency counters:
    (a) Count frequencies   (b) Use analog meters
    (c) Display dc values   (d) Are very large
129. Another instrument that measures frequency is:
    (a) A wattmeter   (b) A frequency multi-
                                        plier
    (c) A frequency meter   (d) A vom
130. Frequency counters can help one to repair:
    (a) Televisions   (b) Radios
    (c) Video games   (d) a, b, and c
131. An oscilloscope uses:
    (a) A digital display   (b) A CRT display
    (c) A D'Arsonval meter   (d) An LCD display
132. Which of the following *is not* an oscilloscope function?
    (a) Automobile engine analyzation
    (b) Stereo amplifier distortion checks
    (c) Waveform measurements
    (d) Signal generation
133. Oscilloscopes:
    (a) Give a visual representation of voltage expressed over a precise
        time interval

    (b) Give a visual representation of persistance expressed over a precise time interval

    (c) Give a visual representation of time expressed over a precise voltage interval

    (d) None of the above

134. Oscilloscopes *cannot*:

    (a) Measure capacitors   (b) Handle complex signals

    (c) Superimpose values   (d) Measure current

135. A capacitance meter:

    (a) Checks capacitor voltage

    (b) Determines the capacity of resistance

    (c) Checks the value of an unmarked capacitor

    (d) Measures watts

136. A wattmeter:

    (a) Determines a circuit's voltage

    (b) Determines power output from a radio transmitter

    (c) Measures dc current

    (d) Is a fictitious device

137. Diodes are temperature sensitive:

    (a) When forward-biased   (b) When reverse-biased

    (c) When new   (d) When held underwater

138. An rf probe is connected to:

    (a) A wattmeter   (b) A capacitance meter

    (c) A D'Arsonval meter   (d) A vom

139. An rf probe *is not* used for:

    (a) Monitoring radio transmitter output

    (b) Repairing radio transmitters

    (c) Adjusting antenna matches

    (d) Measuring dc volts

140. Which of the following components *cannot* be tested by a vom?

    (a) Transistor   (b) Resistor

    (c) IC   (d) Diode

CHAPTER 8

141. Vacuum tubes:

    (a) Don't generate heat   (b) Are compact

    (c) Are semiconductors   (d) Are mostly obsolete

142. The simplest type of semiconductor is the:

    (a) Diode   (b) Triode

    (c) Transistor   (d) IC

143. Diodes:

    (a) Control inductance   (b) Control current direction

    (c) Control leakage   (d) Don't control anything

144. The two most common semiconductor materials are:
    (a) Silicon and argon
    (b) Germanium and silicon
    (c) Germanium and polonium
    (d) Iron and lead
145. Which of the following *is not* a doping substance?
    (a) Boron
    (b) Aluminum
    (c) Gallium
    (d) Iron
146. Two semiconductor material types are:
    (a) p-type and n-type
    (b) a-type and z-type
    (c) 1-type and 2-type
    (d) a-type and b-type
147. Holes behave like:
    (a) Electrons—only positively charged
    (b) Electrons—only negatively charged
    (c) Neutrons—only positively charged
    (d) Atoms
148. A forward-biased diode:
    (a) Has n-type material facing the positive battery terminal
    (b) Has a-type material facing the negative battery terminal
    (c) Has n-type material facing the negative battery terminal
    (d) Has 1-type material facing the negative battery terminal
149. Which of the following *is* a transistor type?
    (a) npn
    (b) azz
    (c) 121
    (d) aba
150. Which of the following *is not* a transistor element?
    (a) Base
    (b) Collector
    (c) Emitter
    (d) Anode
151. Transistors can:
    (a) Amplify voltage
    (b) Amplify current
    (c) Amplify voltage and current
    (d) None of the above
152. Individual components are usually called:
    (a) Discrete components
    (b) Indiscrete components
    (c) Integrated components
    (d) Desert components
153. An integrated circuit contains:
    (a) Many transistors
    (b) Many resistors
    (c) Many capacitors
    (d) a, b, and c
154. A binary number system is based on two digits:
    (a) 1 and 2
    (b) 0 and 1
    (c) 10 and 20
    (d) 1 and 1000
155. A band or dot on the end of a diode indicates:
    (a) The cathode
    (b) The anode
    (c) The emitter
    (d) The plate
156. Today, most components are connected together:
    (a) Through point-to-point wiring
    (b) On printed circuit boards

(c) With glue　　　　　　　　(d) On breadboards

157. A logic probe is used to test:
    (a) High voltage circuits　　(b) 12 volt ac circuits
    (c) 12 volt dc circuits　　　(d) 5 volt dc circuits

158. On an IC, a notch or dot indicates:
    (a) Pin 1　　　　　　　　　(b) Pin 12
    (c) A defective chip　　　　(d) Nothing

159. A CPU is a part of a:
    (a) Microprocessor　　　　(b) Transistor
    (c) Diode　　　　　　　　　(d) Circuit board

160. LSI semiconductors manipulate:
    (a) Hertz of data　　　　　(b) Siemens of data
    (c) Bits of data　　　　　　(d) Bites of data

CHAPTER 9

161. Oscillators:
    (a) Produce ac through a　(b) Eliminate dc
        generator
    (c) Produce ac electron-　(d) Vibrate
        ically

162. Which of the following devices *does not* use an oscillator?
    (a) Radio　　　　　　　　(b) Metal detector
    (c) Music organ　　　　　(d) Lamp

163. Oscillators produce:
    (a) Sine waves　　　　　　(b) Sawtooth waves
    (c) Square waves　　　　　(d) a, b, and c

164. All oscillators must have:
    (a) Feedback　　　　　　　(b) Return waves
    (c) Strobe effects　　　　(d) Antennas

165. All oscillators must also have:
    (a) Amplification　　　　　(b) Deamplification
    (c) Semiconductors　　　　(d) Wire

166. Which of the following components would keep an oscillator's electron flow at a constant level?
    (a) Diode　　　　　　　　(b) Transistor
    (c) Resistor　　　　　　　(d) LED

167. Which of the following circuits *can* be found in an oscillator?
    (a) Bridge rectifier cir-　(b) Filter circuit
        cuit
    (c) Resonant circuit　　　(d) Dc generator circuit

168. Another name for an oscillator's primary inductor is a:
    (a) Switcher coil　　　　　(b) Tickler coil
    (c) Feeder coil　　　　　　(d) Primary system coil

169. The inductor in Question 168:
    (a) Induces energy to replace that lost by resistance
    (b) Induces energy to replace that lost by inductance

(c) Induces energy to replace that lost by oscillation

(d) Induces energy to replace that lost by reluctance

170. Colpitts oscillators use:

(a) No coils
(b) One coil
(c) Many coils
(d) A tapped coil

171. Colpitts oscillators work through:

(a) Capacitive amplification
(b) Capacitive deamplification
(c) Capacitive strobe effect
(d) Capacitive feedback

172. Hartley oscillators are used extensively in:

(a) Automobile engines
(b) Radios
(c) Computer loops
(d) Digital displays

173. Hartley oscillators use:

(a) No coils
(b) Many coils
(c) Much power
(d) A tapped coil

174. A multivibrator produces:

(a) Square waves
(b) Sine waves
(c) Sawtooth waves
(d) A tapped coil

175. A multivibrator is a type of:

(a) Hartley oscillator
(b) Colpitts oscillator
(c) Swirsky oscillator
(d) Resistance-capacitance oscillator

176. What type of wave patterns are used extensively in television receivers?

(a) Square waves
(b) Sine waves
(c) Swinging waves
(d) Triangular waves

177. RC oscillators use:

(a) Inductors
(b) Diodes
(c) Capacitors
(d) Much power

178. RC oscillators employ a pair of:

(a) Diodes
(b) Transistors
(c) Capacitors
(d) Power supplies

179. A differential circuit is added to a:

(a) Multivibrator
(b) Hartley oscillator
(c) Colpitts oscillator
(d) Power supply

180. A differential circuit is used to produce:

(a) Square waves
(b) Sine waves
(c) Sawtooth waves
(d) Triangular waves

CHAPTER 10

181. Radio waves travel at:

(a) 120,000 miles per hour
(b) The speed of sound
(c) The speed of light
(d) 200 miles per second

182. The uppermost radio signals are transmitted at about:

(a) 300,000 Hz   (b) 300,000 kHz
(c) 300,000 MHz   (d) 300,000 GHz

183. The lower the frequency:
(a) The higher the an-tenna   (b) The lower the an-tenna
(c) The shorter the an-tenna   (d) The longer the an-tenna

184. The farthest a radio wave can travel is:
(a) 3,000 miles   (b) 20,000 miles
(c) 200,000 miles   (d) There's no limit

185. On earth, radio signals travel great distances by:
(a) Skipping off the air
(b) Skipping off the troposphere
(c) Skipping off the ionosphere
(d) Skipping off clouds

186. Directional antennas:
(a) Concentrate their signals to one specific area
(b) Spread their signals in many directions
(c) Can be used only to transmit
(d) Can be used only to receive

187. What type of radio wave skips best?
(a) Longwave   (b) Mediumwave
(c) Shortwave   (d) Microwave

188. Satellites operate on:
(a) VLF frequencies   (b) LF frequencies
(c) HF frequencies   (d) vhf and higher frequencies

189. Satellites act as:
(a) Relay stations   (b) Reflect stations
(c) Transmitters only   (d) Receivers only

190. Audio waves:
(a) Are modulated on top of the carrier wave
(b) Are modulated beneath the carrier wave
(c) Are modulated into the carrier wave
(d) Are carried into the modulation wave

191. A diode separates the two waves mentioned in Question 190 through:
(a) Amplification   (b) Inductance
(c) Rectification   (d) Filtering

192. TV's full name is:
(a) Television   (b) Radiotelevision
(c) Imagetelevision   (d) Radiofax

193. Sending still photographs via radio is called:
(a) Radiofacsimile   (b) Radioteletype
(c) Radiopix   (d) Radiophotography

194. Radio in the United States is controlled by the:
(a) FAA   (b) CAB

(c) FCC   (d) USIA

195. People who legally build and experiment with radio transmitters are called:

    (a) CBers   (b) Amateur radio operators

    (c) Electronics hobbyists   (d) HFers

196. What is the comprehensive name for *all* radio frequencies?

    (a) The radio frequency spectrum   (b) The radio service

    (c) The radio frequency band   (d) The radio ether

197. The radio transceiver:

    (a) Is a portable radio   (b) Is a radio beacon

    (c) Is a transmitter/receiver   (d) Is a portable transmitter/receiver

198. Radio waves are:

    (a) Visible   (b) Invisible

    (c) Visible only in water   (d) Visible under ultraviolet light

199. The entry-level radio experimenter's license is called the:

    (a) Novice license   (b) Technician license

    (c) CB license   (d) Beginner's license

200. A field strength meter:

    (a) Measures rf inside the transmitter

    (b) Measures radio frequency voltage at a chosen location

    (c) Measures antenna length

    (d) Measures radio capacitance

SELF-TEST ANSWERS

1.	(d)	17.	(c)	33.	(b)	49.	(c)
2.	(c)	18.	(d)	34.	(a)	50.	(b)
3.	(a)	19.	(a)	35.	(d)	51.	(a)
4.	(b)	20.	(d)	36.	(a)	52.	(a)
5.	(a)	21.	(c)	37.	(b)	53.	(b)
6.	(b)	22.	(a)	38.	(c)	54.	(c)
7.	(a)	23.	(b)	39.	(a)	55.	(b)
8.	(c)	24.	(c)	40.	(b)	56.	(a)
9.	(b)	25.	(a)	41.	(b)	57.	(b)
10.	(a)	26.	(a)	42.	(a)	58.	(c)
11.	(b)	27.	(d)	43.	(a)	59.	(a)
12.	(b)	28.	(b)	44.	(a)	60.	(c)
13.	(b)	29.	(d)	45.	(d)	61.	(b)
14.	(b)	30.	(c)	46.	(d)	62.	(a)
15.	(d)	31.	(d)	47.	(c)	63.	(d)
16.	(d)	32.	(c)	48.	(b)	64.	(a)

65.	(c)	99.	(a)	133.	(a)	167.	(c)
66.	(c)	100.	(a)	134.	(a)	168.	(b)
67.	(d)	101.	(b)	135.	(c)	169.	(a)
68.	(a)	102.	(a)	136.	(b)	170.	(b)
69.	(c)	103.	(b)	137.	(b)	171.	(d)
70.	(b)	104.	(a)	138.	(d)	172.	(b)
71.	(a)	105.	(d)	139.	(d)	173.	(d)
72.	(b)	106.	(c)	140.	(c)	174.	(a)
73.	(b)	107.	(a)	141.	(d)	175.	(d)
74.	(a)	108.	(b)	142.	(a)	176.	(a)
75.	(d)	109.	(d)	143.	(b)	177.	(c)
76.	(a)	110.	(a)	144.	(b)	178.	(b)
77.	(b)	111.	(b)	145.	(d)	179.	(a)
78.	(b)	112.	(c)	146.	(a)	180.	(c)
79.	(d)	113.	(a)	147.	(c)	181.	(c)
80.	(b)	114.	(b)	148.	(c)	182.	(c)
81.	(d)	115.	(b)	149.	(a)	183.	(d)
82.	(b)	116.	(c)	150.	(d)	184.	(d)
83.	(c)	117.	(b)	151.	(c)	185.	(c)
84.	(a)	118.	(a)	152.	(a)	186.	(a)
85.	(d)	119.	(a)	153.	(d)	187.	(c)
86.	(a)	120.	(d)	154.	(b)	188.	(d)
87.	(b)	121.	(d)	155.	(a)	189.	(a)
88.	(a)	122.	(a)	156.	(b)	190.	(a)
89.	(c)	123.	(a)	157.	(d)	191.	(c)
90.	(a)	124.	(a)	158.	(a)	192.	(b)
91.	(b)	125.	(b)	159.	(a)	193.	(a)
92.	(a)	126.	(d)	160.	(c)	194.	(c)
93.	(c)	127.	(b)	161.	(c)	195.	(b)
94.	(b)	128.	(a)	162.	(d)	196.	(a)
95.	(c)	129.	(c)	163.	(d)	197.	(c)
96.	(d)	130.	(d)	164.	(a)	198.	(b)
97.	(d)	131.	(b)	165.	(a)	199.	(a)
98.	(b)	132.	(d)	166.	(b)	200.	(b)

Appendix B

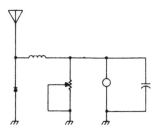

Relative Conductivity of Metals

These are the relative conductivities of common metals at 68°F (20°C). Copper = 100.

Metal	Relative Conductivity
Aluminum	59
Brass	28
Cadmium	19
Chromium	55
Cobalt	16.3
Copper, annealed	100
Copper, hard drawn	89.5
Gold	65
Iron, pure	17.7
Lead	7
Mercury	1.6
Molybdenum	33.2
Nichrome	1.5
Nickel	15
Platinum	15
Silver	106
Steel	7
Tin	13
Tunsten	28.9
Zinc	28.2

Appendix C

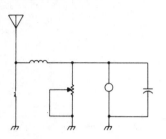

Copper Wire Resistance

This table shows the relationship between copper wire resistance (in ohms) and wire length.

Wire Gauge	Ohms per 1,000 Feet	Feet Per Ohm
10	1	1,000
20	10	100
30	100	10
40	1,000	1

Appendix D

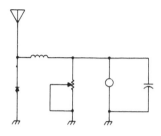

Electrical Safety

When working on circuits (and you should never touch any energized circuit), always make it a habit to keep one hand in your pocket. By doing this, should you encounter some unexpected potential (from an undischarged filter capacitor, perhaps), your chances of accidental electrocution will be greatly reduced. But should one hand touch the chassis, the other hand a voltage source, you will complete a circuit from one arm to the other causing current to flow right through your heart.

To give you an idea of the effects of various current flows on the human body, study the chart below. But, remember, this chart is intended as only a general guideline. Perspiration and other factors can lower body resistance, meaning that as little as .025 amperes of household ac can kill.

Current	Symptom
.05 to 2 mA	Barely noticeable
2 to 10 mA	Slight to strong muscular reaction
10 to 25 mA	Strong shock, inability to move
25 to 50 mA	Violent muscular contractions
50 to 200 mA	Heart irregularities
200 mA and up	Breathing paralysis

Appendix E

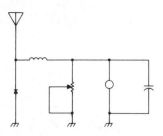

Mail-Order Parts Suppliers

BNF Enterprises, 119 Foster Street, Peabody, MA 01960

Chaney Electronics, Inc., P.O. Box 27038, Denver, CO 80227

Digital Research: Parts, P.O. Box 401247, Garland, TX 75040

Jameco Electronics, 1355 Shoreway Road, Belmont, CA 94002

Poly Paks, Inc., P.O. Box 942, Lynnfield, MA 01940

Quest Electronics, P.O. Box 4430, Santa Clara, CA 95054

Ramsey Electronics, 2575 Baird Road, Penfield, NY 14526

Semiconductors Surplus, 2822 North 32nd Street, #1, Phoenix, AZ 85008

Appendix F

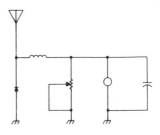

Periodicals

One of the best ways of keeping up on the latest developments in electronics is by subscribing to one or more of the periodicals specializing in your area of interest. Gone are the days when there were only three or four general-interest electronics publications. Today, you'll find dozens of periodicals on all sorts of interesting specializations. Here's a list of some of the most popular. You should be able to examine most of them at any large library.

General-Interest
Computers & Electronics, 1 Park Avenue, New York, NY 10016
Radio-Electronics, 200 Park Avenue South, New York, NY 10003

Computing
Byte, Peterborough, NH 03458
Creative Computing, P.O. Box 789-M, Morristown, NJ 07960
Desktop Computing, Peterborough, NH 03458
Interface Age, P.O. Box 1234, Cerritos, CA 90701
Microcomputing, Peterborough, NH 03458
Personal Computing, Rochelle Park, NJ 07950
Popular Computing, Peterborough, NH 03458

Amateur Radio
Ham Radio Magazine, Greenville, NH 03048
73 Magazine, Peterborough, NH 03048
QST, Newington, CT 06111
CQ, Hicksville, NY 11801

Index

Edited by Roland Phelps